Writing with a Computer

Mike Palmquist
Colorado State University

Donald E. Zimmerman
Colorado State University

Allyn and Bacon
Boston ■ London ■ Toronto ■ Sydney ■ Tokyo ■ Singapore

Vice President, Humanities: Joseph Opiela
Series Editorial Assistant: Rebecca A. E. Ritchey
Executive Marketing Manager: Lisa Kimball
Composition and Prepress Buyer: Linda Cox
Manufacturing Buyer: Suzanne Lareau
Cover Administrator: Jenny Hart
Production Editor: Christopher H. Rawlings
Editorial-Production Service: Omegatype Typography, Inc.
Electronic Composition: Omegatype Typography, Inc.

A Viacom Company
160 Gould Street
Needham Heights, MA 02494

Internet: www.abacon.com

Between the time Website information is gathered and then published, it is not unusual for some sites to have closed. Also, the transcription of URLs can result in typographical errors. The publisher would appreciate notification where these occur so that they may be corrected in subsequent editions.

Library of Congress Cataloging-in-Publication Data

Palmquist, Michael.
Writing with a computer / by Mike Palmquist, Donald E. Zimmerman.
p. cm.
Includes bibliographical references and index.
ISBN 0-205-27487-0
1. Authorship—Data processing. 2. Report writing—Data processing. 3. Word processing. I. Zimmerman, Donald E. II. Title.
PN171.D37P35 1999
808′.02′0285—dc21 98-22324
CIP

Screen image and photograph credits appear on page 268, which constitutes an extension of this copyright page.

Printed in the United States of America

10 9 8 7 6 5 4 3 2 1 03 02 01 00 99 98

For our families

Contents

chapter 10

appendix A

appendix B

appendix C

Preface

Ever since the ancient Sumerians started keeping records on clay tablets five millennia ago, writers have searched for more convenient, efficient, and enduring ways of working with words. Computers—and in particular the word processing software that runs on them—are the latest and most powerful tools available to writers. With a computer, you can easily create, revise, and edit text. You can share your text with others—by printing it, saving it on a diskette, sending it via electronic mail, or publishing it on the Internet. And you can join online communities of writers in order to exchange ideas, find support, and obtain feedback on your latest project.

Given their capabilities, it's not surprising that computers have grown increasingly important in schools and the workplace. Teachers now ask students to do more with technology and employers have come to expect both new hires and long-time employees to have strong writing and computing skills. In fact, in many careers these skills are no longer an option—they're a requirement.

Writing with a Computer addresses the demand for strong writing and computing skills by helping you use your computer to write more effectively and efficiently, to locate and evaluate information more easily, and to better organize information for later use. In the chapters that follow, we call your attention to proven computer-based strategies for enhancing your writing and ability to work with information. We illustrate these strategies using the leading word processing programs on the Macintosh and Windows platforms—and do so in ways that translate easily to the vast majority of word processing programs.

Written in a friendly, easy to understand style, *Writing with a Computer* provides:

- A clear overview of writing processes including generating ideas, drafting, collaborating, revising, editing, and designing documents

- Detailed discussions, with examples and illustrations, of how computers can support specific writing activities
- A comprehensive overview of information resources on the Internet and how to use them
- A focused discussion of strategies for evaluating information found on the Internet
- Guidance on how to find support online, ranging from helpful software to online communities of writers
- Straightforward treatments of advanced topics, including how to customize your operating system and word processing software and how to use many of the latest features offered in word processing programs

For beginning writers, *Writing with a Computer* offers insights into composing processes that will prove valuable with or without a computer. Experienced writers will welcome the clear discussions of how computers can support specific strategies for invention, drafting, revision, collaboration, editing, and manuscript design. Regardless of your writing experience, you'll find that *Writing with a Computer* consistently and carefully creates links between your goals as a writer and the capabilities of your computer.

If you're new to computers, you'll welcome the detailed discussions of the features and capabilities offered by computer operating systems and programs. Key concepts such as the idea of a *desktop* and the differences between *files*, *folders*, and *icons* are clearly and carefully defined. Common features in leading word processing programs are discussed in straightforward terms. And the wide variety of information sources available on the Internet—as well as strategies for using them to support your writing—are explained and illustrated in ways that will help you quickly get up to speed writing with your computer.

If you're an experienced computer user, you can look forward to discussions of how to use advanced capabilities in word processing programs, how to customize your desktop and word processing program, and how to conduct advanced searches of online information sources. You'll also welcome our discussions of ergonomic issues related to writing with a computer.

Most important, you'll find that *Writing with a Computer* provides the necessary guidance to help you build a foundation for continued development of both your writing and computing skills.

OPERATING SYSTEMS AND WORD PROCESSING PROGRAMS

One of the issues we faced in writing this book is how to address differences among leading word processing programs such as Corel, WordPerfect, Lotus Word Pro, and Microsoft Word, as well as operating systems such as the Mac OS, OS/2, and the various versions of Microsoft Windows. We are also concerned about changes that might take place in specific operating systems and word processing programs over the years that readers use this book. Our approach concerning these issues is to consider your general needs as a writer in relation to operating systems and word processing programs focusing on broad principles rather than on specific software commands. Our thinking about this issue can be summed up simply: "Word processing programs change. Operating systems change. But the needs of writers remain largely the same." In *Writing with a Computer*, we focus on your needs as a writer rather on the specific, and changeable, characteristics of operating systems and word processing programs.

This goal has been eased by general similarities among operating systems and word processing software. Most operating systems, as we note in Chapter 7, have *desktops*, *folders*, and *files*. Questions about how you arrange the folders and files on your desktop have more to do with your goals as a writer, it turns out, than with the specific capabilities of an operating system. Similarly, the core capabilities of word processing programs—and even many of the advanced writing tools available in leading programs—not only function in similar ways, but can in many cases be accessed using identical commands.

To keep our focus on your needs as a writer rather than on specific operating systems and word processing programs, we've chosen to refer to word processing programs and operating systems using fairly generic terms. When we talk about specific commands, as we sometimes do in our panels, we call attention to the capabilities of particular word processing programs only when clear differences exist.

HOW TO USE *WRITING WITH A COMPUTER*

You can use *Writing with a Computer* in many different ways. If you are new to computers, work your way through the chapters progressively trying out various techniques and strategies as you read. If you are an experienced computer user, use *Writing with a Computer* as a reference tool. Look up the information you need and try the techniques we suggest. Throughout *Writing with a Computer*, we provide tips to help you write more easily and use your computer more efficiently.

This book consists of two parts. Chapters 1 through 5 focus on how a computer can support your composing process. In Chapter 1, you can read about the processes of composing—generating ideas, collecting information, planning, drafting, organizing, and reviewing and revising documents. Chapter 2 continues the discussion of composing, focusing on editing techniques that can improve a nearly finished document. In Chapter 3, we discuss how you can use your computer to improve the layout and design of documents. In Chapter 4, we discuss techniques for locating, evaluating, and organizing information available online.

Chapter 5, in which we discuss a range of online support available for writers, concludes our focus on composing process issues.

Chapters 6 through 10 address more advanced topics. In Chapter 6, we focus on how computers can help you save time when you write. In Chapter 7, we discuss strategies for organizing a computer desktop in order to make writing easier. Chapter 8 focuses on your larger writing environment, with attention to how you can create a writing environment in your home or office. In Chapter 9, we talk about how you can keep your computer in good working order. Chapter 10 focuses on strategies you can use if your computer stops working properly.

We conclude *Writing with a Computer* with several appendices including a:

- Glossary
- List of common word processing commands
- Discussion of how to work with computers that use the DOS operating system
- List of publications dealing with writing and computers

We hope you'll find *Writing with a Computer* a valuable resource. Many years of work—learning to use new computers, new operating systems, and a wide range of word processing programs and related software—have contributed to our understanding of how best to use computers to support writing. We hope this book will help you write more effectively and efficiently. And, as fellow writers, we wish you well with your endeavors.

ABOUT THE AUTHORS

We bring more than three decades of teaching, research, and writing experience to *Writing with a Computer*. As writing teachers, we have taught a wide range of composition, journalism, technical communication, professional communication, and nonfiction writing classes in a variety of classroom settings. Since the late 1980s, almost all of our classes have been taught in computer-supported classrooms that have allowed students to make extensive use of resources located on the Internet. Mike, a faculty member in the English department at Colorado State University, is a specialist in computer-supported writing instruction and director of the University Composition Program. Don, a professor of technical communication at Colorado State, played a leading role in establishing his department's graduate program in technical communication and served for several years as its director.

As researchers, we've explored a variety of topics related to the use of computers for writing and communication. Our work includes studies of composing processes, reading processes, and document design, as well as studies of classroom instruction. As writers, we bring a wide range of professional communication experience spanning newspaper reporting and editing, freelance writing, the writing of magazine articles, journal articles, and books, multimedia production, and World Wide Web site design and development. In *Writing with a Computer*, we share our knowledge, gathered through this experience, about how computers can help you write more effectively and efficiently.

ACKNOWLEDGMENTS

More than anyone else, we are indebted to our families for the support and patience they showed as we worked on this book. Their support as we've worked on this project is greatly appreciated. We are also thankful for the guidance—and the opportunity to write this book—given to us by our editor, Joe Opiela, and his editorial assistant, Rebecca Ritchey. We are indebted as well to our reviewers who provided thorough and thoughtful comments on early drafts: Dr. Anne Bliss, University of Colorado, Boulder; Douglas Eyman, Cape Fear Community College; Lisa Gerrard, UCLA; Will Hochman, University of Southern Colorado; and Margaret Syverson, University of Texas, Austin.

We would like to extend our thanks to Pat Berlig, physical therapist from the Colorado State University's University Health Services, who reviewed our discussion of ergonomics in Chapter 10. We also want to thank our local technical support staff. Over the years, Kevin Foskin, Kelly Ippolito, and Mike Sueirro, computer specialists in Colorado State University's College of Liberal Arts, have solved more problems with our computers than we care to recall. We appreciate the wide range of troubleshooting strategies they have shared with us as they have solved problems with both our network and personal computers.

Finally, we are grateful to our students and colleagues, who over the years have helped us gain a better understanding of how computers and computer networks can help writers be more productive, better informed, and sometimes even more creative.

chapter 1

Making Your Computer Part of Your Writing Process

Word processing programs are now the tools of choice for most writers. Yet surprisingly few writers invest the time needed to learn more than a fraction of the ways computers can improve their writing.

The amount of time you need to invest is not large. A quick overview of this chapter will provide you with several techniques that can help you generate ideas or develop plans for a document. Further study of the sections in this chapter, coupled with reflection about writing strategies that have and haven't worked well for you on previous writing projects, will provide you with a richer understanding of how computers can help you write.

In this chapter, we turn first to an overview of writing processes, then we discuss specific processes, calling attention to computer-based strategies that can help you carry out each process. The key processes we focus on are generating ideas, collecting information, planning documents, drafting documents, organizing documents, and reviewing and revising documents.

AN OVERVIEW OF WRITING PROCESSES

Some people think writers are born, not made. We couldn't disagree more. The ability to write clearly and confidently emerges only after many hours of practice. Consider Ernest Hemmingway, one of the great writers of the twentieth century. He spent countless hours practicing his craft, endured numerous rejections from publishers before establishing himself as a leading writer, and consumed innumerable gallons of French coffee as he labored patiently over his prose. Was he talented? Absolutely. Was he an overnight success? Absolutely not.

Decades of research on writing clearly indicate that the ability to write well is by no means an inborn talent. It's clear that some writers are significantly better than others, just as some musicians are truly exceptional. If your goals as a writer are somewhat less lofty than a Pulitzer Prize for Literature or a National Book Award, however, you needn't worry that you don't have what it takes. Writing clearly and concisely is something anyone can do—with practice.

Whether you're pursuing a degree or are already well established in your career, we assume you're reading this book to learn how computers can support your writing. This chapter focuses specifically on how computers can support your writing processes. As we point out in greater detail below, writing processes vary according to the writer, audience, document, and discipline. If you're working for a newspaper or writing an assignment for a class, your writing process is likely to be quite different from that of a research scientist or a novelist. For one thing, you're probably working on deadline. For another, you're writing a different kind of document for a different kind of audience. But most important, you're a different person—and it turns out that writing processes are nearly as individual as fingerprints.

Although current understanding of writing processes is based on rhetorical traditions that date to the Ancient Greeks and Chinese, the past few decades have seen rapid growth in our understanding of processes and strategies employed by successful writers. The key points emerging from this research include:

- Writing is not a linear, lock-step process. Instead, writers move constantly from one process to another, sometimes generating text, sometimes revising, sometimes pausing to generate new ideas, and sometimes noting minor editing problems in a text. Typically, writers move back and forth through their text and through their ideas, at one point stopping to focus on a particular passage in a text and at the next point taking time out to rethink their overall approach to the document.
- The major writing processes are generating ideas (also referred to as "invention"), collecting information, planning, drafting, and reviewing (possibly leading to revising and editing). Writers can engage in these processes individually or collaboratively.
- Writers employ a variety of strategies as they engage in writing processes. Think of writing strategies as techniques or tricks of the trade. A common strategy for generating ideas, for instance, is brainstorming. Another is freewriting, which involves writing continuously without worrying about whether you are writing correctly or even making sense.
- Good writers vary their processes according to their purpose and audience. Although good writers may prefer to write in a particular way, they recognize that some occasions require deviating from the norm. If a memo needs to be written within the hour, a good writer will recognize that seeking feedback from a colleague and then revising may be out of the question. Similarly,

writers who like to work alone may find themselves collaborating on a team report. The demands of working with others calls for a much different process than the one they'd follow if they were working on their own.

- Writing processes vary according to how much the writer knows about subject and audience, as well as the kind of document that will be written. If you are writing about a subject that you know well for an audience with whom you are familiar, you are likely to use a different process than if you are writing about a subject and audience with which you are unfamiliar. In the former case, you're likely to spend less time generating ideas, collecting information, and planning and reviewing with attention to your audience. In the latter case, you're likely to spend more time on those processes. Similarly, if you're familiar with the kind of document that you're writing—a lab report, for instance—you'll need to devote less time to understanding the conventions of the document than if you've never written that kind of document before.
- All good writers do not write alike. Although there are certainly common strategies and even some general approaches to writing, a particular writer's process is as unique as his or her handwriting.
- To improve their writing, good writers reflect on their writing processes and strategies. Musicians think about the advantages and disadvantages of particular techniques. Golfers consider the merits of using particular strokes at a given point in a round. Similarly, good writers think about their general approach to writing and how particular strategies can enhance their writing processes. The also consider their preferred style of learning—whether, for instance, they learn best by doing, by listening, by viewing, or by reading. Your learning style can shape whether you begin to write by planning or launch immediately into drafting. It will also shape the way you generate ideas and collect information.

As you consider your writing process and strategies, think about how the computer can help support them. You'll find that many of the commands and writing tools available in your word processing program can help you write more efficiently and effectively.

Panel 1.1
Composing with a Computer

For most writers, typing speed is the biggest obstacle to faster writing. If you're a "hunt and peck" typist, you know what we mean. But even if you type at a respectable speed—say, 30 words or more per minute—a small investment of your time can pay rich dividends. The ability to touch type—to type without needing to look at the keyboard—can increase the speed at which you can create text and free your eyes to do

continued

Panel 1.1 *(continued)*

other things. For instance, you can see the text as it appears on the screen rather than staring at the individual keys as you type them and then reviewing the text you've just typed. Or you can review your notes or other materials as you type, allowing you to more easily translate your notes into text.

Not only are keyboarding skills important for writing, they also have become a prerequisite for many careers. The days are quickly passing, if they aren't already gone, when professionals have secretaries type their documents or word processing clerks enter their documents into computers. As computers have become more common, employers have come to expect that new employees will have good computer skills.

For some careers touch typing skills are even more critical than others. For more than 50 years, journalism departments have required that students demonstrate touch typing proficiency, typically at least 30 words a minute, before taking basic newswriting classes. Long before newspapers, magazines, and publishing houses were computerized, journalists wrote with typewriters, and writing with typewriters enabled them to boost their writing productivity.

How, then, do you develop your keyboarding skills?

Consider taking a keyboarding class or teaching yourself to use a keyboard. Many high schools, junior colleges, business colleges, continuing education, and other education programs offer keyboarding and word processing classes. If you'd like to take a class, check with local schools in your community to see what keyboarding classes they offer. The classes usually run from 10 to 15 weeks and cost between $100 and $200.

Another option is to buy a touch typing software package and teach yourself. Commercial typing tutor programs cost between $25 and $100 for the latest versions and you can, occasionally, find older commercial versions in discount bins and computer supply catalogs. Some software companies provide demo copies that you can download from the World Wide Web at no cost. You can try them to decide whether you want to purchase the full version. Shareware touch typing programs are also available; they're software programs written by individuals who let you download and use the program for free. Shareware works on the honor system. If you like it, you send the author a check for the program.

Before you purchase typing tutor software, make sure it will run on your computer. Most software comes in versions for the Macintosh and IBM PCs. You'll want to make sure you pick up the right version of a program.

To learn the latest programs, search the World Wide Web, or search the online magazine databases for software review articles and obtain them from your local library. Some of the commercial and shareware programs include: Mavis Beacon Teaches Typing, BlackBoard Typing Tutor, Typing Tutor Seven, Mario Teaches Typing, MasterMind Typing, Power Typing, and Typequick. Typing skills programs can provide videos illustrating how to position your hands, wrists, and arms during typing, links to online resources for enhancing your typing skills, and responses as you work through typing exercises.

If you'd like a printed book that teaches touch typing, search the online version of *Books in Print,* available at many libraries, or the printed version of *Books in Print*. It's also available at most book stores.

Once you have a program or a book, plan to spend half an hour a day practicing your touch typing skills. Within a week, you'll be well on your way to learning touch typing.

GENERATING IDEAS

If you've ever stared at a blank piece of paper or a blank computer screen wondering what you were going to write—and when you were going to write it—you know the difficulty of generating good ideas. Sometimes referred to as the process of "invention," a Roman term for a concept developed by the Greeks, writers have long struggled with how to come up with good ideas in a timely fashion. Many of the invention strategies routinely used by writers, in fact, were first used thousands of years ago.

Not surprisingly, the emergence of the computer as the writer's tool of choice has led to significant changes in how writers generate ideas. But the overall goal of coming up with good ideas quickly and efficiently remains the same. Below, you'll find 11 invention strategies that capitalize on the capabilities of computers.

Brainstorming. Most writers have brainstormed at one time or another. Brainstorming often takes place in groups, but you can also do it yourself. If you have access to a network and willing collaborators, you can brainstorm in a group online, possibly in a Chat room (see Chapter 4) or video conference. You can also brainstorm over time using electronic mail or a bulletin board. One advantage of online group brainstorming is the ability to record your discussion. Although recording a video conference may be impractical (since the size of the files holding the video recording will be quite large), recording discussions conducted using Chat, electronic mail, and bulletin boards works quite well. At the end of a Chat session, for instance, you'll have every idea proposed during the discussion. There is no danger of missing a good idea because someone forgot to write it down.

You can also use a computer to record face-to-face brainstorming sessions. A laptop computer, for instance, provides a good way to keep track of ideas generated during a session. After the session, you can share copies of the file containing session notes with group members.

If you brainstorm by yourself, the computer can provide a convenient place to record your ideas. You can open several windows in your word processing program and type related ideas into each window. Or you can simply create a list, then work with the list to identify ideas worth a second look and develop categories as needed.

Avoid self-editing or critiquing during brainstorming. All ideas may not be worthy of further consideration, but wait until the session is over to evaluate ideas.

Freewrite. As the name implies, freewriting involves writing "freely" about a topic or idea. Freewriting extends the idea of avoiding self-editing or evaluating ideas to written text. While brainstorming sessions typically generate lists of ideas, freewriting produces extended text containing elaborated ideas.

Panel 1.2
Switching between and Arranging Windows

Often, you'll find working with more than one document at a time boosts your writing speed. Word processing programs allow you to do this through the Windows menu. You can switch between windows using your mouse, the keyboard, or the menu. And you can display one or more windows at a time (Figure A). Some word processing programs even allow you to view different parts of the same document by "**splitting**" a window (Figure B). A split window is particularly useful when comparing different parts of a document—for example, comparing an overview with a section.

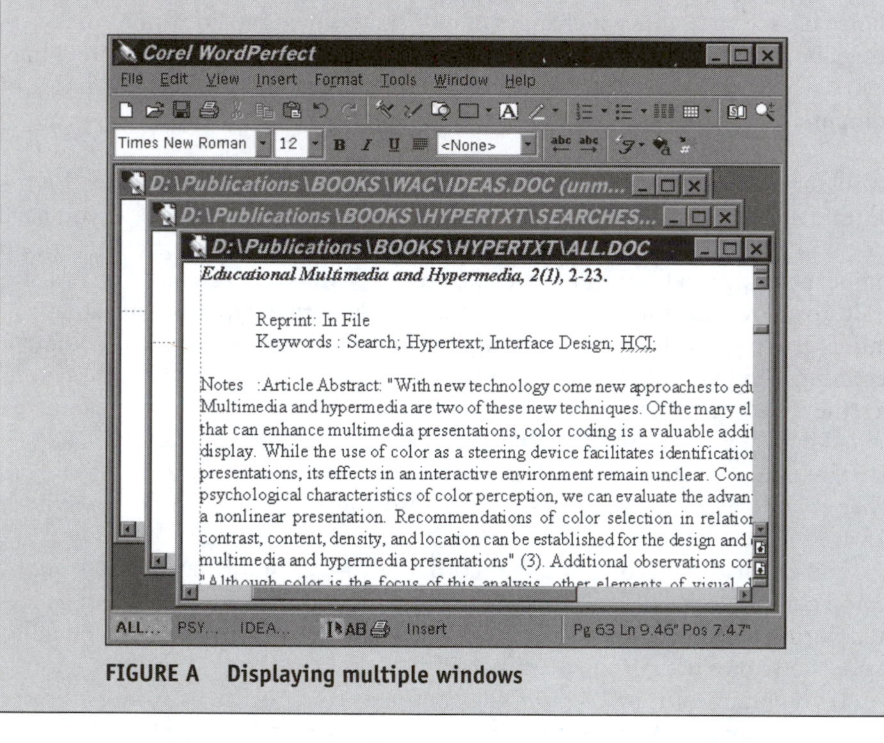

FIGURE A Displaying multiple windows

The computer lends itself well to freewriting. Simply open a new document in your word processing program and start writing. Some writers set a timer for 10 or 15 minutes and write without stopping until the timer sounds—even if they're only writing, "I'm not coming up with an idea yet." Freewriting is not a foolproof way of generating worthwhile ideas, but it works surprisingly well.

Blind Write. Blind writing, a technique uniquely suited to the computer, is a variation on freewriting. Writers who find it difficult to avoid self-editing while they freewrite benefit from this technique. To blind write, open a new document

Panel 1.2 *(continued)*

Writers use multiple windows and split windows for many reasons. If you are working on a document that builds on a previous document, you can borrow text from the earlier document. If you are working on related documents, you can use windows to ensure you're describing things in the same way. Or if you are writing a report, you can display your notes in one window and work on your report in another.

HOW TO SPLIT A WINDOW

1. Click on the Windows or View menu
2. Click on SPLIT

Note: In Microsoft Word, you can also split a window by pulling down the SPLIT WINDOW button at the top of the Scroll Bar.

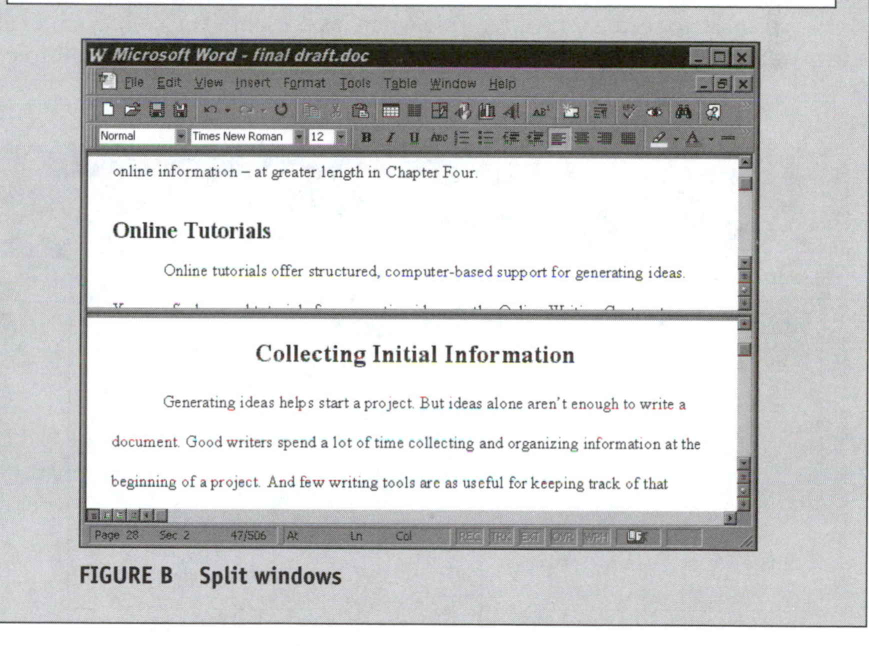

FIGURE B Split windows

in your word processing program, turn off or turn down your computer monitor, and start freewriting. Since you won't be able to see what you're writing, it won't be as easy to self-edit or critique your text. As with freewriting, setting a timer and writing until it sounds can be useful.

Looping. Looping also builds on freewriting. After writing for 5 or 10 minutes, reread what you've written and highlight the most important idea or passage.

Copy the text you've highlighted into a new document and freewrite again. Then repeat the process. Looping allows you to generate and refine an idea so that you can decide whether it's worth considering or rejecting as the basis for the document you want to write.

Comparison Tables. Tables can help you explore ideas more easily than you can using normal text. Sometimes you can see relationships between key ideas more clearly. Sometimes you can see whether sufficient support exists to support a proposal.

Use the INSERT TABLE or CREATE TABLE command to create comparison tables (Panel 1.3). Your table can take many forms. Use a three-column table to explore the advantages and disadvantages of a particular idea. Use a two-column table to generate pro and con arguments for a position. Use tables to compare information drawn from various sources and to contrast positions taken by various authors on a position (Figure 1.1).

In addition to seeing text differently when it's presented in a table, tables also allow you to easily reorganize information. You can move text from cell to cell.

Question: Should the U.S. adopt a national school voucher initiative?			
Author	Position	Admitted Pros	Admitted Cons
Garcia	Yes, as soon as possible	Increased competition leads to better schools Better than current system	Some bugs will need to be worked out of system initially; there are dangers of elite schools still being out of reach of the poor
Johnson	Not nationally, it's a local decision; but is in favor of vouchers	Increased competition leads to better schools Can foster renewed debate about goals of our educational system	Not a national, but rather a local issue Not certain how system will work in all parts of the country
Alexander	No, it's a step toward elite schools for the wealthy and second or third rate schools for the middle class and working poor	Debate has led to frank discussion about merits of current educational system	Will further gap between haves and have nots
Yin	No, but we need to change the system as soon as possible	Current voucher systems in U.S. have suggested some potential benefits	Voucher systems have not yet proven a viable alternative to traditional system

FIGURE 1.1 Author comparison table

Panel 1.3
Inserting Tables

Tables are used in a wide range of documents. As a result, word processing programs offer tools that make it easy to create and work with tables. Leading word processing programs allow you to click on an icon in a button bar and insert a table containing a specific number of preformatted rows and columns (Figure A). Similarly, leading word processing programs make it easy to adjust the width of columns and the height of rows, to delete or merge cells, to add new rows and columns, and to specify cell shading and borders. In addition, text within tables can be formatted like other text.

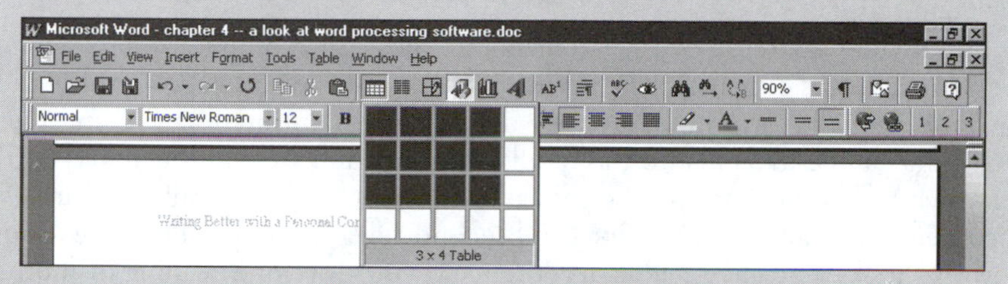

FIGURE A Inserting a table

Tables present information clearly and concisely. They also help show relationships among information. As you work with tables, experiment with different layouts to determine which ones work best with the material you want to place in the table. Consider as well how other writers have used tables in their documents. If you are following a style guide, consult the guide for formatting guidelines.

HOW TO CREATE A TABLE

Using the Menu

1. Place your cursor at the point in the text where you want to create a table
2. Click on the Table, Insert, or Create menu
3. Click on Table
4. Specify Rows and Columns in the Create Table dialogue box
5. Click on Create

Using the Button Bar

1. Place your cursor at the point in the text where you want to create a table
2. Click and hold down on the TABLE icon in the button bar
3. Drag until you have selected the number of rows and columns that you want
4. Release the mouse button

You can reorder columns and rows to explore relationships between ideas. And you can see what tables look like when you delete passages of text.

Tables are useful not only for generating and elaborating ideas, but also for exploring relationships among ideas.

Outlining. Like the other strategies in this section, the decision to use an outline is up to you. Many writers prefer not to use outlines under any circumstances—it's simply not part of the way they write. We use outlines relatively rarely, but have found them to be particularly useful on longer documents. We used an outline on this book both to generate ideas and, later, to review and edit our manuscript.

If you choose to use outlines to generate ideas, the computer offers a number of advantages over paper. Word processing programs offer powerful outlining tools that allow you to easily view documents in outline form. Using these tools, you can easily generate and modify outlines and view as much or as little text as you want. For instance, to gain a better sense of the overall structure of a document, you can view only first- and second-level headings. This gives you a view that looks much like a table of contents. If you move one heading to another part of the document, all text under that heading moves with it.

In addition to allowing you to explore and modify documents, outlining tools allow you to see relationships among ideas. If you begin writing a document by creating an outline, for instance, you can easily rearrange and add new sections before writing any sections. Viewing a document in outline form also allows you to see whether you'll need to add more detail (in the form of subheadings) to a particular section and whether you are using consistent organizational strategies throughout the document.

Even if you are working with short documents, consider using the outlining tools available in your word processing program. Since you can easily arrange and rearrange text, you'll be able to experiment with alternative organizational patterns and explore relationships among key ideas in a document.

Review Old Documents. When faced with a new writing assignment, consider browsing through documents you've written before. Sometimes referred to as "mining" text, writers use this technique to pull new ideas and organizational principles out of old documents. As you read the document, jot down ideas as they come to you or copy key passages from the old document and paste them into a new one.

Consider using old documents not only to generate ideas, but also as organizational and format templates for new documents. Remember that documents need not be completely original to fulfill their purposes. Documents that worked well in the past can provide valuable guidance, and save valuable time, as you begin a new document.

Panel 1.4
Viewing Documents

In leading word processing programs, you can view a document in several ways. You can view it according to page or layout, which shows the document approximately as it will appear when it is printed (Figure A). You can view it in draft or normal view, which shows the document without margins and page breaks (Figure B). Or you can view the document using outlining tools (Figure C). Outlining tools work well if you have assigned heading levels to your text. If you are working with a long document, you can use the outlining tools to **COLLAPSE** or **EXPAND** various parts of your document.

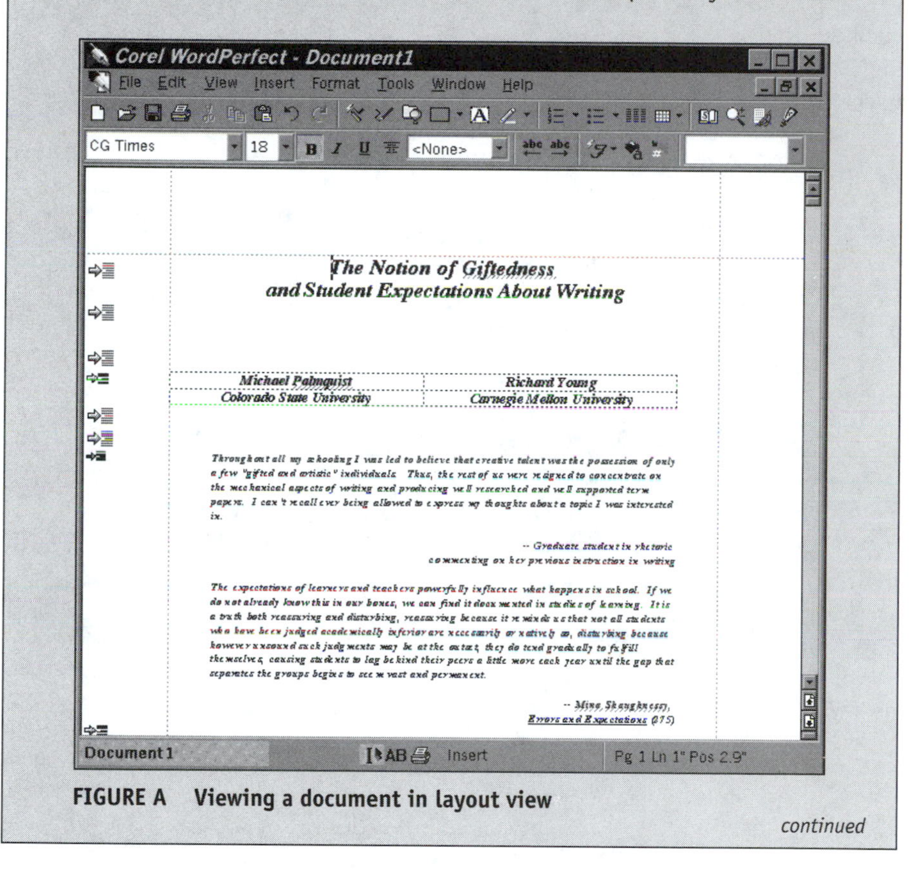

FIGURE A **Viewing a document in layout view**

continued

Browse the Web. Browsing the Web builds on the same principle as reviewing old documents. Web pages can help you see how other writers have treated a topic. They can also alert you to recent developments on the topic. Although you need to be cautious about material on the Web, consider conducting a quick search on a topic when other information-gathering techniques haven't proved

Panel 1.4 *(continued)*

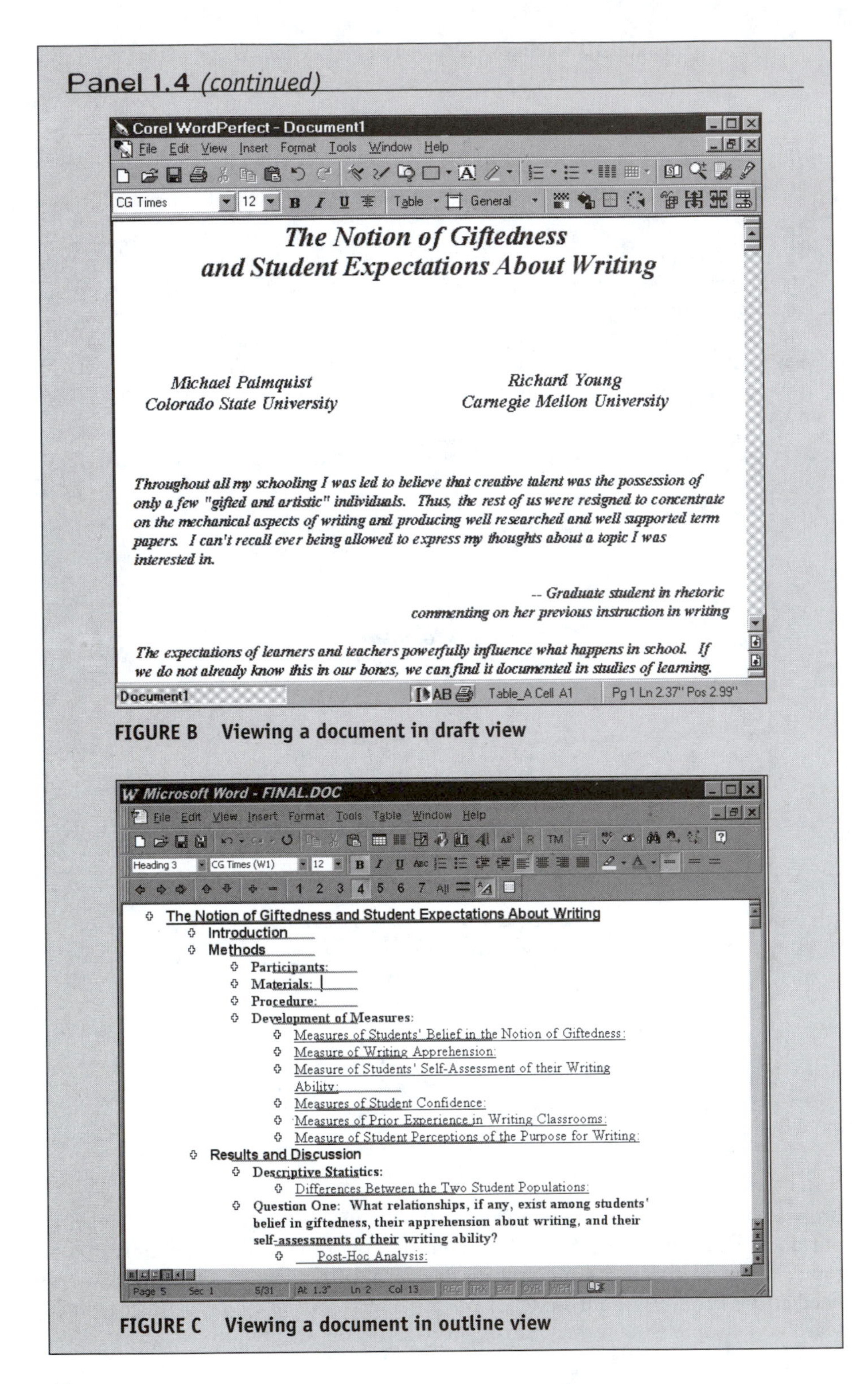

FIGURE B Viewing a document in draft view

FIGURE C Viewing a document in outline view

to be successful. We discuss Web searches and other sources of online information at greater length in Chapter 4.

Online Tutorials. Online tutorials offer structured, computer-based support for generating ideas. You can find several tutorials for generating ideas on the Online Writing Center at Colorado State University <http://www.colostate.edu/Depts/WritingCenter>. The tutorials ask you to respond in writing to prompts. At the end of a tutorial, the program organizes your response and allows you to review and revise it. You can print your response, send it to yourself in an electronic mail message, or copy and paste it into a word processing file. You can read more about tutorials in Chapter 5.

Collaborating. Writers often find that the best way to generate ideas is to talk a project over with colleagues by walking down the hall or calling someone on the phone. After the conversation, they'll list their ideas or freewrite on their computer. These conversations can be more formal, as well. Some writers meet regularly to discuss their writing. In corporate or academic settings, these meetings sometimes use groupware—software that helps groups work together. Many groupware programs use Chat or videoconferencing to support online discussions. Some groupware programs also provide "whiteboards," which allow writers to work collaboratively on documents even when they're separated by thousands of miles. Groupware helps bring writers together, which can be the single most important way to generate ideas for a document.

Sending electronic mail can also help bring writers together. You can ask, "What do you think?" to gain a different perspective on an issue and to learn about new ideas on a topic.

If you'd rather not interact directly with another writer, consider reading a newsgroup, mailing list, or Web-based discussion forum. Reading messages posted to a newsgroup or forum serves purposes similar to browsing old documents or browsing the Web. If you come across a posted message that interests you, consider corresponding via electronic mail with the person who posted the message. You can read more about newsgroups, mailing lists, and Web forums in Chapter 4.

Waiting for Inspiration (a.k.a., the Deadline). It happens to all of us. If we wait too long for inspiration to strike, we run up against a deadline. Some writers thrive on deadlines. The thrill (or is it fear?) of writing against a deadline brings out the best in them. Some writers find that they can't write well, in fact, unless they are facing a deadline.

If you're working collaboratively, avoid this approach to generating ideas. Your colleagues won't enjoy working on your schedule if they must rush about at the last minute completing tasks that might have been done days or even weeks earlier.

On the other hand, if you're working alone, you may find that nothing helps generate ideas like a fast-approaching deadline. If so, find other ways to enjoy life before the deadline approaches. You'll find you become quite busy in the hours before a project is due.

tip 1.1 ■ Save and Back Up Your Documents Periodically

On occasion, the electrical power shuts down, hard drives crash, floppy diskettes become corrupted, and other gremlins strike. If you develop a practice of saving and backing up your files of your documents, you'll recover from such disasters with a minimal loss of time. Consider saving and backing up your file every 15 minutes or so.

To minimize problems:

1. Write down the page number of your document
2. Move your cursor to the top of your document
3. Use the **SAVE** command to save your document files to the C: or hard drive
4. Use the **SAVE AS** command to back up your document to another hard drive or floppy diskette
5. Use the **SAVE AS** command to save your document to your hard drive again under its original subdirectory and same name
6. Under the Edit menu, use the **GO TO** command to return to the page where you were working
7. Continue working on the document

As you work on different versions of your document, save them under different names.

COLLECTING INITIAL INFORMATION

Generating ideas helps start a project. But ideas alone aren't enough to write a document. Good writers spend a lot of time collecting and organizing information at the beginning of a project. And few writing tools are as useful for keeping track of that information as a computer.

In Chapters 4 and 7, we discuss how to organize information obtained during online searches and strategies for organizing files and folders, respectively. In this section, we discuss 10 strategies you can use as a writer to support your writing process. These strategies—which many experienced writers use regularly—can help you set up your computer as a repository of information that you can draw on as needed.

Keep a Journal. Many writers keep journals. Many more think journals are kept only by poets or essayists who support themselves by working in coffee shops. But journals are kept by writers of all kinds: students, corporate CEOs, scientists, successful entrepreneurs, college professors, engineers, and middle managers, among many others. These journals can include detailed notes about scientific research, discussions of strategies for sales campaigns, reflections on conversations, and explorations of new ideas for writing projects.

Computers offer many options for journal writers. Some writers choose to keep a running journal divided according to days. A single file might contain journal entries for a particular month or year. Other writers use individual files for each journal entry, relying on descriptive names and folders to organize their journal.

If you decide to use a journal, you'll be on your way to creating a valuable resource that you can draw upon when you start a new writing project.

Jot Down Notes. If you're not convinced that keeping a journal will work for you, use the computer to store your notes. Some graduate students, for instance, keep a running list of notes about a thesis or dissertation in a file called "musings." A "musings" file allows you to jot down ideas as they occur to you.

Other writers store notes in a personal information manager, such as Lotus Organizer, Microsoft Outlook, or Corel InfoCentral. You can attach notes to entries in an address book, keep track of scheduled meetings, or add items to a to-do list. If you develop a systematic way of keeping track of your notes, they can be a valuable resource on many writing projects.

Store and Organize Electronic Mail Messages. Many writers have found that electronic mail not only simplifies communication, but also provides a convenient way to keep records of activities and projects. Writers in many settings communicate frequently via electronic mail. Messages can cover issues ranging from the minutes of a particular meeting to ideas about strategies to pursue on a collaborative writing project to explanations of how a particular problem was addressed. You can organize your electronic mail by creating folders, which serve the same purpose in electronic mail as they do in your file cabinet, and storing relevant messages in them (see Chapter 7 for a discussion of folders and desktops). When you begin work on a particular writing project, look over your messages for relevant information.

Save Useful Posts to Newsgroups, Mailing Lists, and Web Forums. If you read newsgroups, visit Web forums, or subscribe to mailing lists, you'll run across messages that you might find useful later. Just as you can save and organize your electronic mail messages, you can save messages from newsgroups, mailing lists, and Web forums. We discuss these communication tools in detail in Chapter 4.

Create Lists of Useful Web Sites. Web browsers, such as Netscape Navigator and Microsoft Internet Explorer, allow you to record and organize Internet addresses for Web sites (Figure 1.2). These addresses, called URLs (or Uniform Resource Locators), are referred to as bookmarks (Navigator) and favorites (Internet Explorer). Bookmarks and favorites can be organized in folders that can be accessed from menus or button bars in the programs. You can also add descriptions to each bookmark or favorite to help you recall the contents of each site.

If you spend a lot of time on the Web, bookmarks and favorites offer significant advantages over conducting searches for information. Rather than conducting a search and examining the results of the search one site at a time, you can browse through your bookmarks or favorites to quickly locate relevant sites.

Save Transcripts of Chat Sessions. If you engage in Chat sessions to brainstorm, to discuss progress on projects, or to discuss related issues, you may find it useful to save session transcripts. Just as all electronic mail messages are not

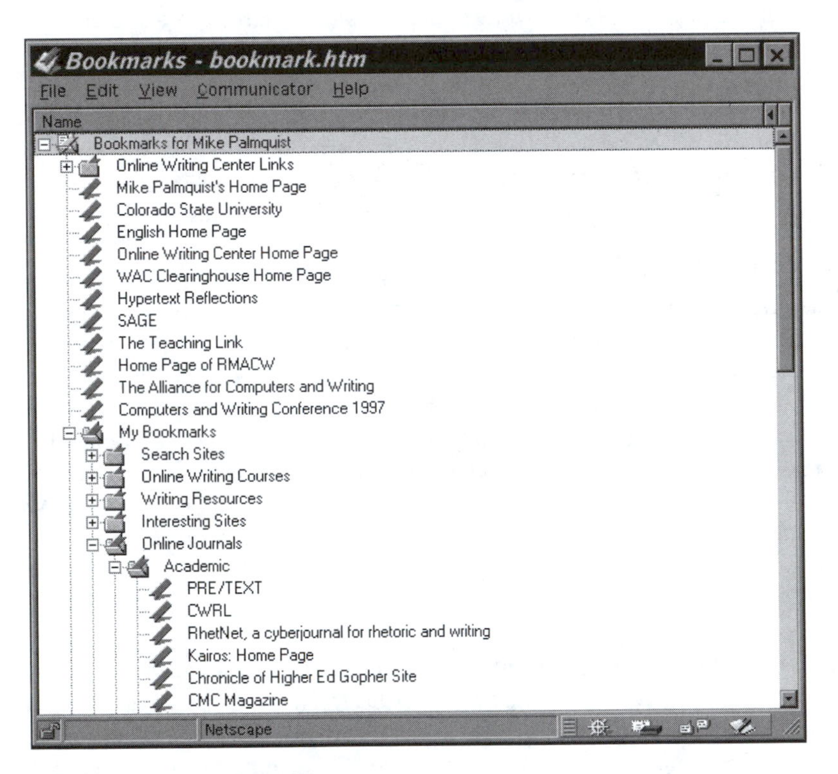

FIGURE 1.2 Bookmarks

worth archiving, all Chat sessions are not worth preserving. Occasionally, however, you'll want to record a session. Since most Chat sessions include a lot of extraneous information, consider excerpting useful passages and discarding the rest.

Store Sample Documents. Old documents are a great source of information for many projects. Many writers—particularly those in academic or corporate settings—find that their writing projects frequently build on previous work. If you make a habit of keeping old documents on your computer, consult them when you begin work on a related project. Use old documents to generate new ideas, to provide formatting and organizational frameworks for new documents, and to provide text that you can reuse.

Collect Boilerplate. "Boilerplate" refers to text that you can use over and over again. A description of a particular institution or department, for instance, serves as boilerplate. There's little reason to come up with a new way of describing an institution or department if it hasn't changed since you last described it.

You can lift boilerplate out of your old documents, or save specific passages into individual files and use them as needed. The latter approach works well when you find yourself regularly using boilerplate. Some writers even create macros (see Chapter 6) that automatically insert passages of boilerplate into documents.

Keep Records of Meetings. If you find yourself writing about decisions made during meetings, keep records of the meetings. When you need to write about the decisions in a report or other document, consult your records. It's helpful to have a member of the group take notes and electronic mail them to everyone who attended the meeting.

You can organize meeting records in folders and consult them as needed during your writing process.

Maintain Bibliographic Databases. The print analog to Web bookmarks and favorites is the annotated bibliography, a list of sources that are briefly described, usually using no more than one or two sentences. If you work extensively with print sources, use a bibliographic database to keep records of what you've read (see Chapter 5). Bibliographic database programs allow you to record publishing information about a print document (author, title, publisher, etc.) and to annotate the document. Leading programs support keywords, which help you search for documents on particular subjects.

If you're writing for academic or trade journals that require works cited lists and in-text citations, you can also use bibliographic database programs to generate works cited lists and automatically insert citations into documents created with most leading word processors.

PLANNING DOCUMENTS

Most writers plan their documents whether they know it or not. Planning involves considering your goals and your strategies for meeting them. Writing researchers distinguish between global planning and local planning. Global planning, as the name suggests, focuses on larger issues of audience, purpose, organization, and style. Local planning focuses on issues such as word choice, sentence structure, and appropriate evidence for a particular point.

When you're thinking about how to make a sentence sound just right, you're engaged in local planning. But concerns about "sounding just right" are based on a larger plan that considers how to best address an audience to meet a particular goal or set of goals. In this way, global plans shape local plans. But the reverse can also be true. If you're finding it difficult to adopt an appropriate voice or style for a document, problems that emerge from local planning—for instance, how to phrase a sentence in just the right way—can lead you to reconsider your global plans.

Research indicates that experienced writers spend much more time planning their documents than do inexperienced writers. Experienced writers also spend a greater proportion of their time addressing global issues, while inexperienced writers do proportionally more planning at the local level.

Why plan when you could save time by just getting right to work? In the long run, planning saves time and effort. Experience suggests the most effective documents emerge from careful consideration of 10 global issues:

- Your audience(s)
- Your purpose(s) for writing a document
- Your role as writer—sole writer, lead writer, or member of a team
- Your strengths as a writer
- Writing strengths that your collaborators, if any, bring to the project
- Your experience writing similar documents
- Your knowledge of the topic
- Your access to information on the topic
- Document constraints such as deadlines, required formats, and length limitations
- Your support for completing the document, for example, secretarial support, research support, printing and graphics support

Don't assume that all planning takes place before you begin working on a document, however. Like other writing processes, planning is strategic, and strategies vary widely. Some experienced writers like to begin drafting for a while, just to get a sense of what they know about a topic, and then stop and plan. Others plan and draft in parallel, allowing an overall plan to emerge as they draft and adapting what they've already drafted to meet that plan. Still others take a structured approach, developing detailed outlines and carefully considering issues of audience and purpose before a single sentence is written.

As important as planning is, few computer software tools support planning. You can, of course, use outlining tools to examine and reflect on the overall structure of a document. You can also explore the structures of related documents and compare them to the current document. And you can use multiple windows to jot down notes about a document for later consideration.

In general, however, you'll find that the most important tools to support planning are the judgment and experience of you and your colleagues.

DRAFTING DOCUMENTS

One of the most misleading myths about writing is that it's simply the process of putting thoughts on paper (or on screen). The writing process, in this view, is simply one of translating fully-formed thoughts into fully-formed sentences and paragraphs. Presumably, you shouldn't start writing until you've figured out what you're going to say.

Writers who buy into this myth usually run into trouble. They grow frustrated with the difficulty of finding the right words to express their ideas. They discover that their organization does not work out as well as expected. They realize that they hadn't fully thought out how their audience was likely to react to what they've written. As a deadline approaches, they realize that the document they thought they could draft quickly after lunch might take until midnight.

Why is drafting such a slow process? Several factors contribute to the difficulty of writing quickly and clearly. As we've suggested, drafting is not a straightforward process of putting your thoughts down on paper. Instead, it's a process of creating text that will help others understand ideas and information.

Research indicates that documents can be written quickly only when a writer is intimately familiar with the subject matter and understands the audience, purpose, and nature of the document being written. Even then, experienced writers frequently stop to read and reread their text, adjust the wording of passages, and move blocks of text to different parts of the document. The drafting of even the most experienced writers, it appears, is anything but the fluid translation of thought into text.

Our observations of experienced writers in protocol studies and usability testing sessions indicate that writers spend more time reading and reflecting about what they've written than they do drafting. The process of reading what's been written, reflecting on what's been read, and revising as needed is repeated throughout the writing process. These reflections often lead writers to take long breaks from drafting to reconsider how best to address their audience, to explore new organizational frameworks for the document, or to collect new information that they suddenly realize they need.

In short, if any one process is at the center of writing, it's drafting. Drafting new text is the best way to find out what you know and don't know about a topic. Below, we discuss six strategies for drafting text using a word processing program.

Recycling. The simplest way to use a computer to support your writing process is to reuse old documents or text passages. If you're writing a letter of recommendation, for instance, open a letter you've written for someone else and use the word processing program's SAVE AS command to save it as a new file. Modify text that addresses the qualifications of the person you're writing about and retain standard passages such as the salutation and closing. Although this strategy can be misused to the point that you change little more than the names in a letter of recommendation, it can save a great deal of time if you write numerous letters.

Recycling works well for larger documents, as well. Use your earlier reports and evaluations as the basis for new documents; they will provide an overall organizational framework for a new document and serve as a source of boilerplate.

Perfect Draft. If you are writing a new draft about a subject you know a great deal about, consider this strategy. The perfect draft strategy is based on the assumption that writers can begin with the first sentence of a document and write straight through the document.

This strategy works quite well if you have a good understanding of your audience, purpose, content, and organizational framework. Given this level of understanding, it can be the most efficient and effective way to write a document. Simply open a new document and begin drafting.

If, however, you're not absolutely sure that you understand all aspects of your writing task, this strategy may not be effective. In some cases, in fact, writers who begin a writing project with the perfect draft strategy end up rethinking their project and discarding much of the text they've written.

If you're confident that you have the necessary knowledge of audience, purpose, content, and organization to write a document straight through from start to finish, consider the perfect draft approach. If you're not confident of these things, consider other drafting strategies.

Successive Refinement. The successive refinement strategy resembles the perfect draft strategy. Start at the beginning of the text and write as much as possible until reaching a stopping point. At that point, consider why you've hit a stopping point, refine your plans, return to the beginning of the document, and start revising. When you finish revising, return to the end of the draft and begin drafting again.

Successively refining a draft works well when you have a good idea of your general approach and a clear idea of how you want to start. Writers who use this approach read and reread their evolving draft to get a sense of what needs to be revised and what needs to be added to the text.

Like the perfect draft strategy, you need only open a new file and begin writing to carry out this strategy.

Build on an Outline. If you've created an outline for your document, this strategy works well. Use the outlining tools on your word processing program to expand and collapse sections as you work on them individually (Figure 1.3), or view the outline in one window while writing sections in another (Figure 1.4).

Unlike the perfect draft and successive refinement strategies, this strategy doesn't require beginning at a particular point in the draft. Work from the beginning to the end of the document. Or work on individual sections in any order. As your plans change, adapt the outline to your needs. Ideally, you won't need to discard text that you've written. But you may find it necessary to add, combine, or reorder sections.

Bits and Pieces. Many writers prefer to approach a new writing task without an overall plan or outline. Using freewriting and other idea-generating strategies, they write bits and pieces of the document, often in no discernible order. A writer of a scientific report, for instance, might start with a passage describing the

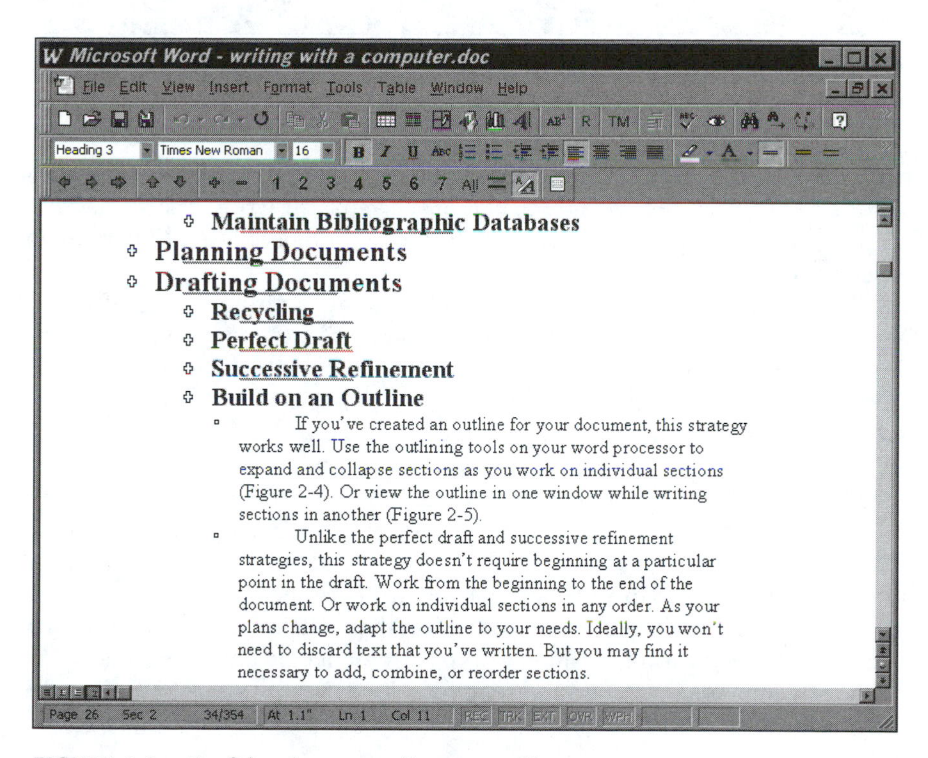

FIGURE 1.3 Drafting by expanding an outline

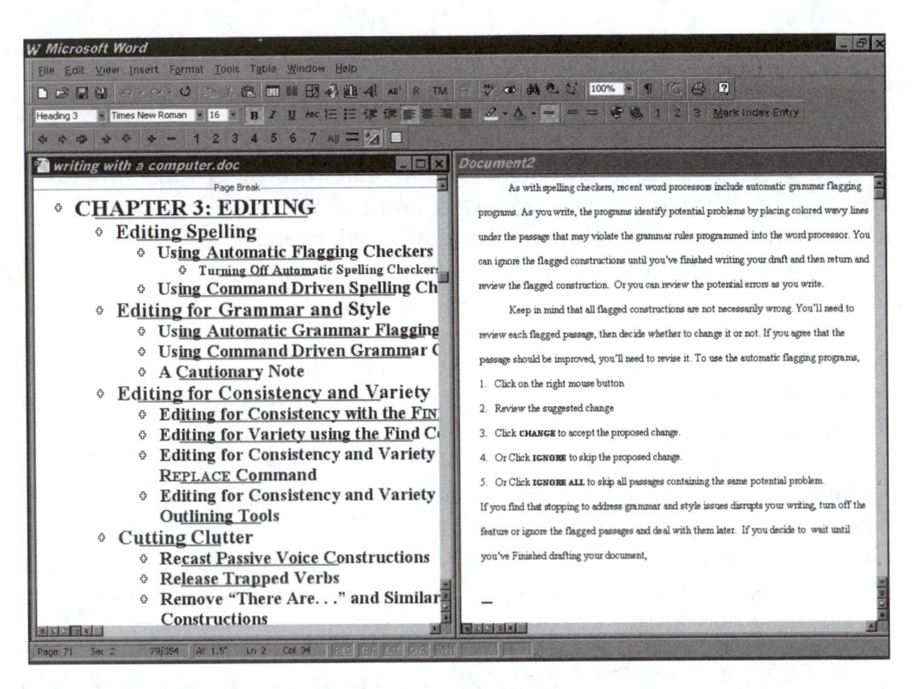

FIGURE 1.4 Drafting by expanding an outline

participants in the study, then shift to the conclusion, then write part of the introduction or literature review. As each new idea occurs, the writer drafts a new passage of text. Some passages may be only a few sentences long. Others might consist of dozens of pages.

The computer is well suited to supporting this drafting strategy. You can write in several windows, saving each to a separate file. Or you can jump around the text in a single document, possibly using a split-window view of the document. However you choose to write individual passages of text, you can use the computer to move easily from one passage to another.

Collaborative Writing. Among writers, the notion of collaboration has grown in importance over the past few decades. The once dominant view of the writer as a solitary being has given way to a more social view of writing, leading writers and writing teachers to explore how collaboration can be supported throughout the writing process. We don't want to suggest that collaborative writing is a new idea. It's long been part of common workplace practices in academia, journalism, technical communication, and industry. We do want to suggest, however, that growing numbers of writers now recognize collaboration's role in many (if not all) forms of writing.

Collaboration can occur at any point in the writing process. Collaborative idea-generating activities are common, for instance, as are collaborative reviews of documents. Drafting can also be a collaborative activity:

- Writers can collaborate by assigning each person to write one or more sections and then merging the sections into one document. It's a common strategy when co-writers bring different strengths to a project. In this book, for instance, we wrote separate chapters that reflected our different experiences and expertise in writing and teaching writing. We each took primary responsibility for drafting our respective chapters, then edited them collaboratively for voice and style. The computer network allowed us to exchange documents via electronic mail and shared network folders. Our word processing programs allowed us to track changes using Version and Document Comparison tools, and to comment on our drafts by making annotations (Figure 1.5).
- Writers can collaborate in a round-robin fashion, with one writer reviewing and revising sections drafted by another. Writers can pass sections back and forth until they are satisfied with the overall result. Computer support for this kind of collaborative drafting is similar to that used to support the

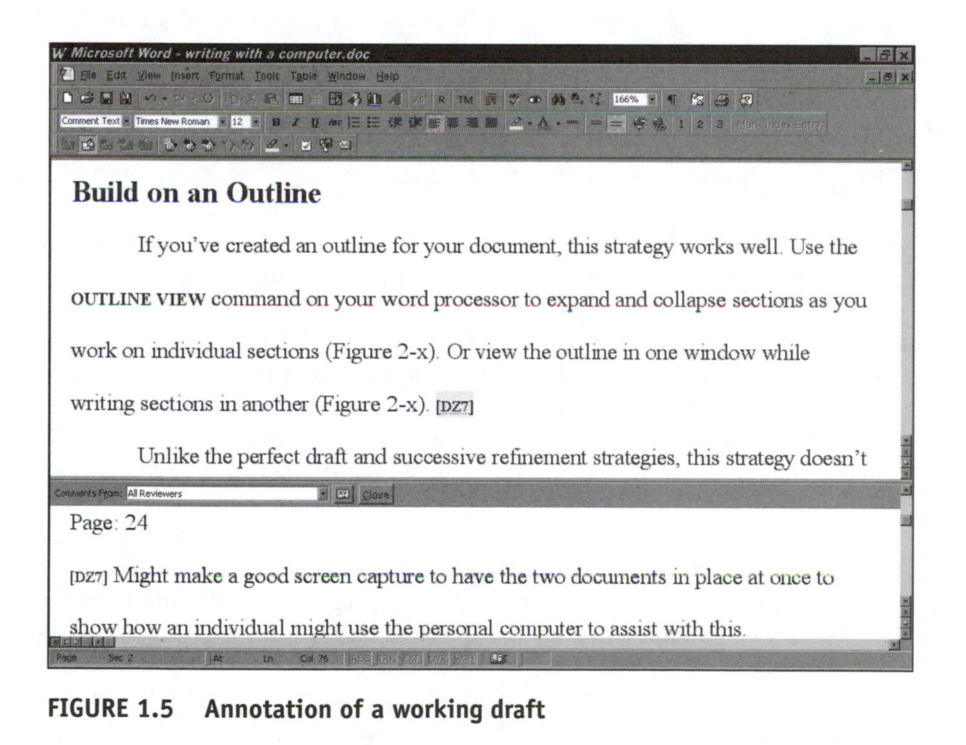

FIGURE 1.5 Annotation of a working draft

Panel 1.5
Tracking Changes in Documents

Word processing programs have grown increasingly sophisticated in their ability to help writers track changes in documents. Two of the most powerful tools for tracking changes are comparison and revision tools and version tools.

Comparison and Revision tools allow you to identify differences between one draft of a document and another. Word processing programs that support comparison and revision marks compare two files—each containing different drafts of the same document—and mark changes in one of the files. Comparison and revision marks are particularly useful if you ask another writer to make changes directly to a document, and later on want to find out what changes were made.

Most word processing programs that support comparison of documents do so on a paragraph by paragraph basis. Paragraphs containing changes are marked, leaving you to identify precisely what has been changed.

Comparison and revision marks are useful if it's been a while since you last reviewed a document. When several days or even weeks have passed before the document is returned to you, you may find it difficult to determine what has or hasn't been changed. You can also use comparison and revision marks when you aren't sure which of two or more files contains the draft you want to work on.

Version tools allow you to examine the history of changes to a text. They are particularly useful when you encounter the all too common problem of "disappearing text." A colleague once told us that he writes on paper instead of on a computer because of this problem. He felt that word processing programs make it too easy to delete text. Once the text is deleted, it's gone forever.

previous strategy. Files can be exchanged on floppy diskette or via computer network. Word processing programs support comparing documents, tracking changes, and annotating documents.

- Writers can collaborate on documents by making different contributions to a document. One writer, for instance, might draft the overall document. Another might add charts and data analysis. Another might edit the document for consistent style and voice. Yet another might add citations and works cited pages and obtain permissions for use of published text or images in the document. The use of networks to share documents during this kind of collaboration is crucial, as is the use of annotations (e.g., "Put the chart here" or "I'm not sure I follow you on this point. Can you elaborate?").
- Some writers prefer to collaborate by engaging in team writing sessions. These can be conducted in a face-to-face setting often by passing the keyboard from writer to writer, or over the network, possibly using groupware to support their interaction. Although this collaborative approach does not work for all writers, those who use this strategy typically find the exchange

Panel 1.5 *(continued)*

Until recently, the disappearing text phenomenon was a fact of life for writers who work on personal computers. Like other writers, we've found ourselves regretting a decision to delete passages of text that, at the time we deleted it, seemed to be of absolutely no use.

To avoid this problem, some writers have saved multiple drafts of a document. In any given folder, as a result, they would have files named something like "Project 1–Draft 1," "Project 1–Draft 2," "Project 1–Draft 3," and so on. Although this approach allows you to keep updates of your drafts, it doesn't solve the problem of text that was deleted between periodic updates.

Versioning, a technique for addressing the problem of disappearing text that has long been used in some mainframe and mini-computer operating systems, has recently been implemented in leading word processing programs for personal computers. Versioning allows you to automatically save a "version" of the document each time you save the document. These versions can subsequently be opened and saved as separate files and changes can be viewed by comparing the documents. To access the version tools in your word processing program, click on the file menu and then click on **VERSIONS** or **VERSION CONTROL.**

Versioning has some limitations. It increases the size of a file and it only works if you decide to turn it on. It also doesn't allow you to restore every change made to a document. If you save a document, write new text, and then delete the text before you save the document again, you won't be able to restore the deleted text.

Nevertheless, versioning helps solve the problem of "lost text." Consider using it if you find yourself wishing that you could review, and possibly restore, changes made to your documents.

conducive to generating ideas and considering alternatives that might not occur to a writer working alone.

- Some writers take a hierarchical approach to collaboration. Using a word processing program's annotation and versioning tools, for instance, the primary writer of a document can seek feedback from reviewers. The reviewers might be content area experts from other parts of an organization. Some might be the writer's supervisors. Others might be writers who are the primary writers for other documents. In each case, the computer can speed collaboration by supporting file exchange and annotation.

ORGANIZING DOCUMENTS

Organizing, like the other processes we discuss in this chapter, can occur at any point, and often at several points, during your writing process. Word processing programs support organizing and reorganizing documents by making it easy to move text from one part of a document to another and by allowing writers to view a document in different ways. Word processing programs also support organization in less obvious ways. You can use a word processing program to generate a table of contents, for instance, that provides you and your readers with an overall view of a lengthy document. And formatting commands, in particular the STYLE command (see Chapter 3), allow you to add a consistent look and feel to a document. Remember, though, that the use of styles or formatting commands are not themselves an organizational technique so much as they are a reflection of an organizational structure that you develop for a document.

Like planning, drafting, generating ideas, and collecting information, the process of organizing a document is influenced by:

- Your audience's expectations regarding document content and format
- Whether you are required to use a particular style and format in your document
- Your purpose for writing the document
- The kind of information you are using in your document

Organization is typically thought of in fairly conventional ways. Many essays and reports use techniques such as cause–effect or compare–contrast as organizational devices. Others use chronology as an organizational principle. Still other documents follow established organizational patterns for a particular genre, such as the scientific report's introduction and review of literature, methods, results and discussion, and conclusion.

Regardless of the organizational pattern they follow, writers who use computers typically follow one of four organizational strategies:

1. You can work within established guidelines for a particular kind of document. This situation is common when you are writing a conventional document such as a term paper, a lab report, or an article for a scholarly journal, or are following company or agency guidelines for a specific type of report. Using a pre-established organization is also common if you are writing a senior design project or a grant proposal.
2. You can build on a previous document. If you are writing a report or a term paper, for instance, consider using a report or paper you recently wrote as an organizational template. Assuming your work was well received, there's little need to depart from a proven organizational pattern. Similarly, consider using old memos or letters as organizational templates for new memos and letters. Writers often put a great deal of thought into organizing documents. If you can use proven organizational patterns—and there's no downside to repeating past practices—you'll save time and effort.
3. You can establish the organizational pattern before drafting. Writers who use this strategy use their knowledge of a document's purpose, audience, and content to develop an overall organizational pattern. Then they fill in the

document in ways consistent with the organizational pattern. Coupled with a drafting process that uses an outlining strategy, this approach allows writers to generate text for specific sections of a document in any order they wish.

4. You can develop an organizational pattern as you draft. This strategy works well in cases where you are drafting text in a more or less random order. As text is drafted, an organizational pattern begins to emerge.

These organizational strategies move from the familiar (or at least conventional) to patterns that can be completely new. Regardless of the strategy you use, the computer offers a range of support for organizing documents. These strategies range from techniques that shape the overall organization of a document, to techniques useful within sections and subsections, to formatting strategies that inform the reader about the purpose of specific text passages.

- Use outlining tools to view the overall structure of your documents. If you use styles to create hierarchical headings (e.g., first-level heading, second-level heading, third-level heading, etc.), you can expand and collapse your view of the document. Word processing programs that support outlining allow you to view as many as 10 levels of detail.
- Use the SPLIT WINDOW command to view different parts of your document at the same time (Panel 1.2). You can compare an overview of a section, for example, with the section itself. This helps you determine whether you've presented your points in the section in the same order that you introduced them in the section overview. You can also use the SPLIT WINDOW command to view different sections that follow (or should follow) the same organizational principle.
- Use charts, graphs, and tables to consolidate information. Charts, graphs, and tables are useful for presenting information that does not lend itself to narrative. Readers usually find it difficult to read series of statistical results in paragraph form, for instance. Presenting the same information in a chart, graph, or table can better direct a reader to the information and help them understand it more easily. We discuss charts, graphs, and tables in greater detail in Chapter 3.
- Use your word processor to create tables of contents and indexes. These organizational devices are of great value to your readers, but you can use them to help you gain a detailed sense of a document's organization. We discuss creating tables of contents and indexes in Chapter 6.
- Call attention to specific information in a document (and create a consistent look and feel for a document) by using the STYLES command, bulleted and numbered lists, and character formatting commands such as bold, italic, and underline. You can also use paragraph indents, line justification, line spacing, icons, colors, and borders to highlight key information and call attention to passages that serve particular functions. We discuss these issues in greater detail in Chapter 3.

REVIEWING AND REVISING DOCUMENTS

Reviewing and revising are similar to the editing processes that we discuss in the next chapter. Both sets of processes involve noting problems in a document, deciding how best to address the problems, and rewriting the document. Reviewing and revising, however, usually focus on larger issues than editing, which focuses on changes at the level of words and sentences. Primary areas of concern during reviewing and revising include:

- How well a document addresses its audience or audiences
- How well a document meets its purpose or purposes
- Whether document organization and format help meet the writer's goals for the document
- Whether effective and appropriate support is provided for key points
- Whether the content is accurate
- Whether issues and ideas in the document are treated consistently

In this section, we'll discuss strategies you can use by yourself and with other writers. We'll focus on reviewing and revising your own documents, reviewing and revising documents written by others, and reviewing and revising documents collaboratively.

Reviewing and Revising Your Own Documents

When you work with your own documents, reviewing and revising may seem like a single process. But you're actually doing three things:

1. You're reading your own work to find any global or local issues that need to be addressed. As you read your document, for example, you might detect inconsistencies in the way you address your audience, or you might note that you haven't offered sufficient evidence to support a key point.
2. You're figuring out how best to address a problem. If you've addressed your audience in different ways throughout a document, you'll most likely decide that you need to adopt a more consistent approach. The key question, of course, is which approach to choose. Similarly, if you've found that a key point needs more support, you'll need to decide what kind of support is needed and where it can be found.
3. You're making changes to your document. After you've decided how to address a problem, you need to change the text in your document to reflect your decision.

Like all the processes that we discuss, review and revision can occur throughout your composing process. Once you've written even a few words, you can begin reading and reflecting on them.

Given the range of concerns writers can address during reviewing and revising, you can use many strategies. Consider the 16 strategies in the following discussion that use the computer to help you review and revise your documents.

Print the Document and Review It on Paper. The first step in any review process is reading the document. You may prefer to print a hard copy of the document for a number of different reasons. You may find it difficult to read text on a computer monitor. You may prefer to spread a document out on your desk in order to get a better sense of its overall structure. Or you may prefer to make annotations in the margins of a printed document. Regardless of the reason, if you prefer to read a document on paper, by all means print it out.

Panel 1.6
Copyediting Symbols

To speed editing printed copy, print a double-spaced copy of your document and then use the copyediting symbols given below.

COPYEDITING SYMBOLS

Meaning	Example
Delete	To reduce the "high" accident rate
Circle means spell out	He deleted (25) pages.
Circle means abbreviate	The (department) of state . . . (abbr. dept.)
Underline three times to indicate capitalization	Sally and jack fell into the river.
Draw a slanted line to change to lower case	Donna and Sam swam across the River.
Delete and close up	The policy-holder bought a new policy.
Underline once to indicate italics	Elements of Style remains a useful book.
Begin new paragraph	¶ Sally dropped the phone as she reached to turn on her computer.

continued

Panel 1.6 *(continued)*

Meaning	**Example**
Center text	]Chapter 22[
Set flush left	[Bibliography
Set flush right	Page 26
Insert missing letter	Armadillos are comon in Oklahoma. (m inserted)
Insert missing word	Armadillos have ^ shells. (hard)
Transpose letters	Most suknks have a strong odor.
Separate words	Most writers work#slowly#on longer documents.
Insert period	John bought screws, nuts, and bolts⊙
Insert comma	. . . pencils, folders^ and books.
Insert question mark	What have you done^?
Insert hyphen	. . . the black^footed ferrett.
Let original text stand	Sally divided ~~her the~~ large job . . .
Close up text	For our own project consider breaking the job into five tasks: (1) identify problem, (2) propose solutions, (3) select preferred method, (4) assign personnel, and (5) implement management plan.
Mark out grossly misspelled copy and write new copy above	John could not determine the word misspelled ~~smippseelled~~.

Review a Document in Page View or Normal View. Some writers prefer to review documents on their computers. Using the PAGE VIEW or LAYOUT VIEW command, you can read the document as it will appear when printed. Using the NORMAL VIEW or DRAFT VIEW command, you can read the document without reference to how it will appear when it is printed. If you have a large computer monitor (17 inches or larger) and a high resolution display, you may find it as easy to read text online as it is on paper. In addition, you'll reduce your use of paper.

Review and Revise Document Organization with Outlining Tools. Outlining tools can help you review and revise a document's overall structure. Assuming that headings in the document have been formatted using the STYLE command, you can view a document in increasing levels of detail (Panel 1.4). You can, for instance, gain an overall sense of the document's organization by viewing first- and second-level headings. Or you can view lower-level headings to gain a more detailed understanding of the document's structure.

Using outlining tools, you can also assess the organizational patterns used in sections of your document. You can quickly determine, for instance, if you are following similar organizational patterns in each section. You can also assess the effectiveness of the organizational patterns you've chosen.

Outlining tools also allow you to assess the support you've provided for your key points. By collapsing and expanding sections of your document, you can identify the amount and kind of support you've used in each section. You can determine, for instance, whether you've provided comparable amounts of support for each of your key points. You can also determine how effectively you have used particular pieces of information, such as quotes or statistics, to support your points.

Finally, outlining tools allow you to easily reorganize sections of a document. To move a section to a new location, drag the section with your mouse or use the CUT and PASTE commands (see Appendix B for more information about common editing commands).

Review Document Structure and Format Using the ZOOM Command. The ZOOM command also allows you to assess the overall organization of your document (Panel 1.7). You can zoom out in PAGE or LAYOUT VIEW to view several pages at once. This allows you to determine how you've organized specific sections of a document, for example. It also allows you to quickly assess how much text you've devoted to each section.

The ZOOM command also allows you to move easily around your document. You can place your cursor on a particular page, zoom in to view the page in full size, and then zoom back out to see the document as a set of thumbnail pages.

Highlight Support for Key Points. Leading word processing programs include a HIGHLIGHT command, which allows you to highlight text in colors of your

Panel 1.7

ZOOM

ZOOM refers to the relative size of the page on the computer screen (Figure A). You can **ZOOM** pages so that they appear much larger than they would appear on paper—a useful technique if you are working on detailed graphics or tables. Or you can **ZOOM** pages so that they appear much smaller than they would appear on paper—a useful technique if you want to view the placement of graphics on a page.

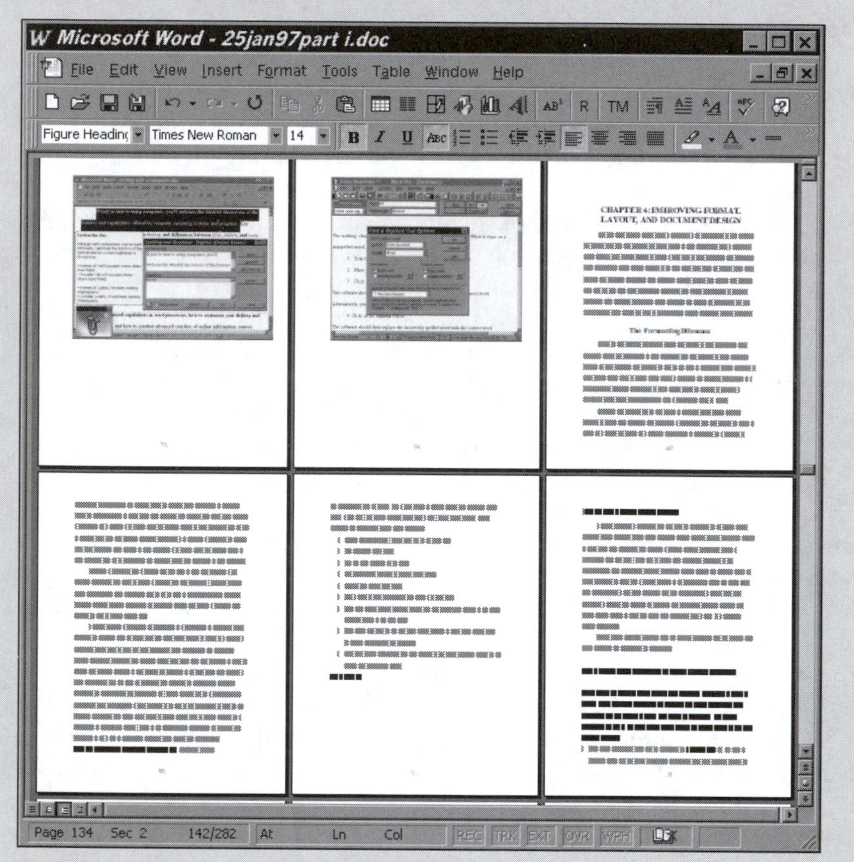

FIGURE A Viewing multiple pages using ZOOM

HOW TO USE ZOOM

Using the Menu

1. Click on the View menu
2. Click on **ZOOM**

Using the Button Bar

1. Click on the **ZOOM** icon on the button bar

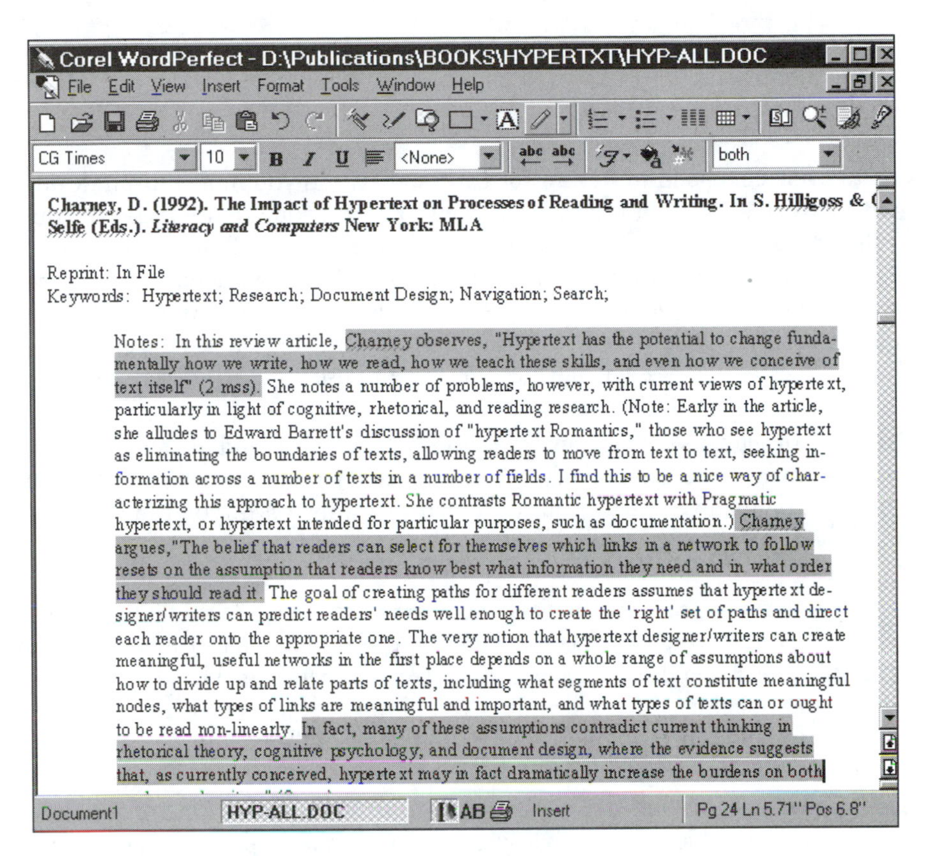

Charney, D. (1992). The Impact of Hypertext on Processes of Reading and Writing. In S. Hilligoss & C. Selfe (Eds.). *Literacy and Computers* New York: MLA

Reprint: In File
Keywords: Hypertext; Research; Document Design; Navigation; Search;

Notes: In this review article, Charney observes, "Hypertext has the potential to change fundamentally how we write, how we read, how we teach these skills, and even how we conceive of text itself" (2 mss). She notes a number of problems, however, with current views of hypertext, particularly in light of cognitive, rhetorical, and reading research. (Note: Early in the article, she alludes to Edward Barrett's discussion of "hypertext Romantics," those who see hypertext as eliminating the boundaries of texts, allowing readers to move from text to text, seeking information across a number of texts in a number of fields. I find this to be a nice way of characterizing this approach to hypertext. She contrasts Romantic hypertext with Pragmatic hypertext, or hypertext intended for particular purposes, such as documentation.) Charney argues,"The belief that readers can select for themselves which links in a network to follow resets on the assumption that readers know best what information they need and in what order they should read it. The goal of creating paths for different readers assumes that hypertext designer/writers can predict readers' needs well enough to create the 'right' set of paths and direct each reader onto the appropriate one. The very notion that hypertext designer/writers can create meaningful, useful networks in the first place depends on a whole range of assumptions about how to divide up and relate parts of texts, including what segments of text constitute meaningful nodes, what types of links are meaningful and important, and what types of texts can or ought to be read non-linearly. In fact, many of these assumptions contradict current thinking in rhetorical theory, cognitive psychology, and document design, where the evidence suggests that, as currently conceived, hypertext may in fact dramatically increase the burdens on both

FIGURE 1.6 Reviewing by highlighting a draft

choosing (Figure 1.6). Use the HIGHLIGHT command to identify the support you've provided for each of your key points. Use different colored highlights to identify different kinds of support or to identify the source of your support. For, example, you can highlight all support from published studies using the color yellow and all support from field sources in blue. Using this approach, you can determine whether you are relying more heavily on one kind of support than another and whether you are using a similar mix of support in each part of your document.

Highlight Passages for Subsequent Review and Revision. You can also use the HIGHLIGHT command to mark a text for subsequent revision. You can use different colors to indicate the need for different kinds of revision. Green might indicate that the passage needs additional support. Red might indicate that a passage needs to be revised to better address your audience. Blue might indicate problems with organization.

Used in combination with the ZOOM command, this strategy allows you to quickly identify the kinds and amount of revision you need to engage in after you've completed your review.

Use the FIND Command to Review for Consistency. If you find yourself dealing with a particular concept in several sections of a text, you can use the FIND command to locate each point in the text at which you address the concept (Panel 1.8). As you move through your document, you can determine whether you've dealt with the concept consistently. This strategy also works well when you need to use standard descriptions in a document. You can use the FIND command to locate each passage in which you need to use a specific description.

Use the FIND AND REPLACE Command to Revise Text. If you find inconsistencies in your treatment or description of particular concepts or objects, use the FIND AND REPLACE command to revise your document. In some cases, you can substitute entire passages using this command (Panel 1.8).

Use the SELECT, DRAG, COPY, CUT, and PASTE Commands to Revise Text. The SELECT, DRAG, COPY, CUT, and PASTE commands allow you to move text in a document or to replace existing text with new text. These commands enable you to relocate sections in a document. They also allow you to replace old text with new text. We discuss these and other common editing commands in Appendix B.

Use the SAVE AS Command to Save Drafts of Your Document. One of the drawbacks of revising on a computer is losing text. In most word processing programs, you cannot recover deleted text once you've closed a file. To reduce the danger of losing text, you can periodically SAVE drafts of your document. When you begin a major revision of a document, for instance, save the document to a new file under a new name and revise the new file. If you need to consult your original draft, you'll still have it in a separate file.

Use the Versions Tool to Keep Track of Changes to Your Document. If your word processing program supports a Versions tool (Panel 1.5), you can use it to keep track of changes to a document. Versions works much like the SAVE AS command, but does not require you to save your document to a new file. If you need to track changes in a document, you can open a version for review or save a specific version of the file to a new file.

Review Documents Using Document Comparison Tools. Document comparison tools (Panel 1.5) allow you to identify differences between two documents. You can review a document by comparing it to an earlier draft and noting changes between drafts.

Panel 1.8

FIND and REPLACE

The FIND (or SEARCH) command allows you to locate specific letter combinations, words, or phrases within a text (Figure A). FIND is useful if you realize you've consistently misspelled a word, overused a modifier, or used a phrase in a way that, on reflection, you wish you hadn't. Typically, word processing programs allow you to indicate whether case—capitalization or the lack thereof—should be considered in the search. Many word processing programs also allow wild card symbols in a search (typically asterisks or question marks), which allow you to locate words containing variant spellings or multiple endings.

FIGURE A Using FIND

The REPLACE command allows you to search for a specific text string (a letter or letters, one or more words, or a phrase) in a text and substitute them with another string of text (Figure B). Like the FIND command, you can usually use wild card symbols to search for text that you would like to replace.

FIGURE B Using REPLACE

continued

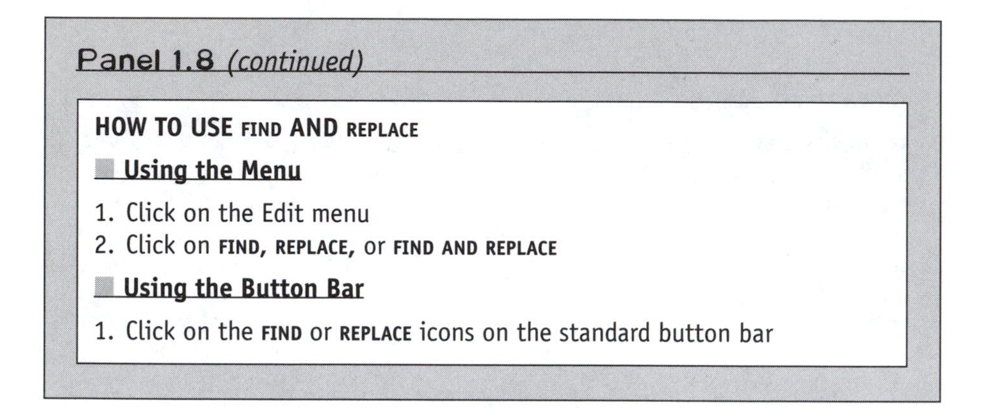

Panel 1.8 *(continued)*

HOW TO USE FIND AND REPLACE

Using the Menu

1. Click on the Edit menu
2. Click on FIND, REPLACE, or FIND AND REPLACE

Using the Button Bar

1. Click on the FIND or REPLACE icons on the standard button bar

Use Multiple Windows to Compare Drafts of a Document or to Compare Related Documents. If you've saved drafts of a document to different files, you can compare the documents using multiple windows (Panel 1.2). You can view the windows one at a time, moving between them using the **WINDOW** command or the mouse, or you can arrange the windows so that they display side by side.

You can also use multiple windows to compare related documents. If you are working on a report similar to one you completed earlier, for example, you can compare the two documents to determine whether you can borrow text or organizational patterns from the earlier document.

Record Your Ideas in Annotations, Footnotes, or Endnotes. Consider beginning your document review by reading it quickly without making revisions. This provides you with an overall sense of the document and reduces time wasted on revisions that subsequent reading will tell you are unnecessary. When you review a print document, you can easily make comments in the margin; likewise, when you review an online document, you can make similar comments in annotations, footnotes, or endnotes (Panel 1.9). The advantage of this strategy is that you can either hide or view these annotations, footnotes, and endnotes. You can also move directly from an annotation, footnote, or endnote to the passage that you commented upon.

Record Your Ideas in the Comments Field in the Document Properties Dialogue Box. If you are using a word processing program that provides a Document Properties dialogue box, you can use the dialogue box to record your ideas (Panel 1.9). The advantage of this strategy is that all your comments are recorded in a single place and are stored in the same file containing the document.

Record Your Ideas in Another Document. Many writers record their comments on a document in a separate file. You can use the **WINDOWS** command to move

Panel 1.9

Commenting on Texts

Leading word processing programs allow writers and reviewers to comment electronically on a document in several ways. Following the long-established practice of making marginal comments on a document, you can use annotation tools to insert comments within a text. You can also use endnotes and footnotes to add comments to a text. And in some word processing programs, you can make comments in the document's Properties dialogue box.

Word processing programs support annotations in several ways, but all follow the same principle. A reviewer selects text or inserts the mouse cursor at a particular point in the document, opens an annotation window, and writes a comment. The annotation shows up in the document (or in the margin of the document) as an icon or set of highlighted initials indicating that a comment has been made (Figure A). By clicking on the icon or set of initials, the writer can view the comment (Figure B).

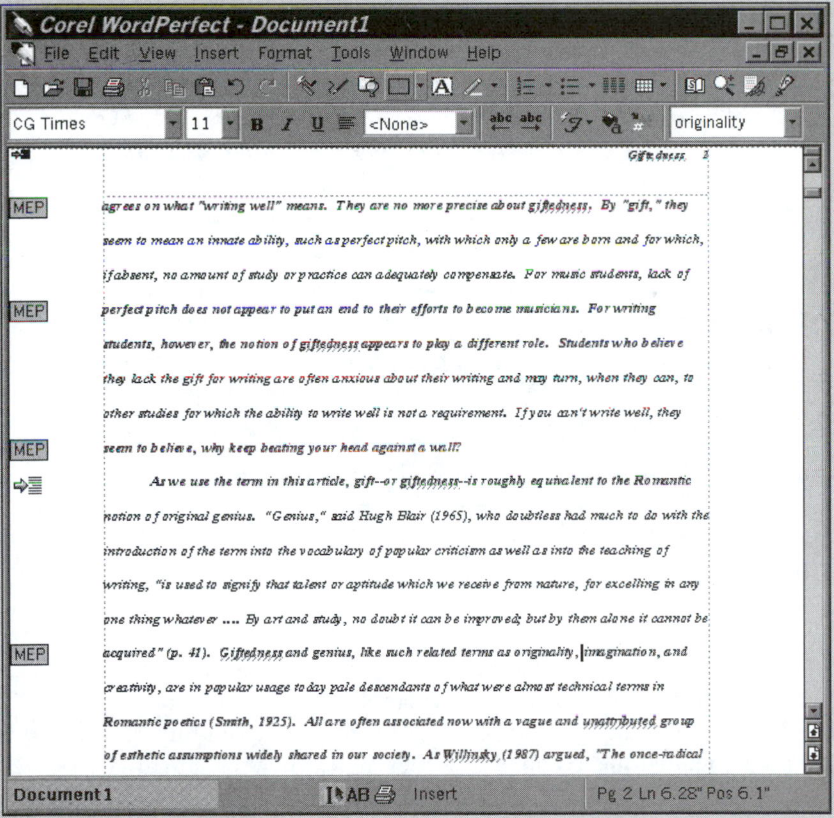

Corel WordPerfect - Document1

File Edit View Insert Format Tools Window Help

CG Times 11 <None> originality

Giftedness 2

MEP agrees on what "writing well" means. They are no more precise about giftedness. By "gift," they seem to mean an innate ability, such as perfect pitch, with which only a few are born and for which, if absent, no amount of study or practice can adequately compensate. For music students, lack of MEP perfect pitch does not appear to put an end to their efforts to become musicians. For writing students, however, the notion of giftedness appears to play a different role. Students who believe they lack the gift for writing are often anxious about their writing and may turn, when they can, to other studies for which the ability to write well is not a requirement. If you can't write well, they MEP seem to believe, why keep beating your head against a wall?

As we use the term in this article, gift--or giftedness--is roughly equivalent to the Romantic notion of original genius. "Genius," said Hugh Blair (1965), who doubtless had much to do with the introduction of the term into the vocabulary of popular criticism as well as into the teaching of writing, "is used to signify that talent or aptitude which we receive from nature, for excelling in any one thing whatever .... By art and study, no doubt it can be improved; but by them alone it cannot be MEP acquired" (p. 41). Giftedness and genius, like such related terms as originality, imagination, and creativity, are in popular usage today pale descendants of what were almost technical terms in Romantic poetics (Smith, 1925). All are often associated now with a vague and unattributed group of esthetic assumptions widely shared in our society. As Willinsky (1987) argued, "The once-radical

Document1 Insert Pg 2 Ln 6.28" Pos 6.1"

FIGURE A Annotation marks

continued

Panel 1.9 *(continued)*

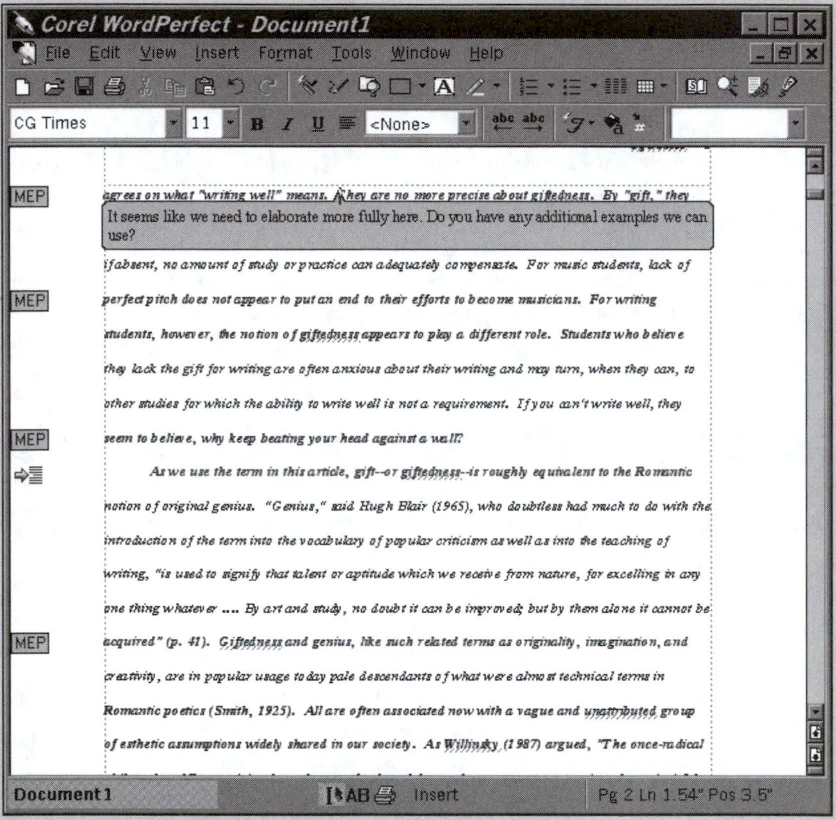

Corel WordPerfect - Document1

File Edit View Insert Format Tools Window Help

agrees on what "writing well" means. They are no more precise about giftedness. By "gift," they

It seems like we need to elaborate more fully here. Do you have any additional examples we can use?

if absent, no amount of study or practice can adequately compensate. For music students, lack of perfect pitch does not appear to put an end to their efforts to become musicians. For writing students, however, the notion of giftedness appears to play a different role. Students who believe they lack the gift for writing are often anxious about their writing and may turn, when they can, to other studies for which the ability to write well is not a requirement. If you can't write well, they seem to believe, why keep beating your head against a wall?

As we use the term in this article, gift--or giftedness--is roughly equivalent to the Romantic notion of original genius. "Genius," said Hugh Blair (1965), who doubtless had much to do with the introduction of the term into the vocabulary of popular criticism as well as into the teaching of writing, "is used to signify that talent or aptitude which we receive from nature, for excelling in any one thing whatever .... By art and study, no doubt it can be improved; but by them alone it cannot be acquired" (p. 41). Giftedness and genius, like such related terms as originality, imagination, and creativity, are in popular usage today pale descendants of what were almost technical terms in Romantic poetics (Smith, 1925). All are often associated now with a vague and unattributed group of esthetic assumptions widely shared in our society. As Willinsky (1987) argued, "The once-radical

Document1 Insert Pg 2 Ln 1.54" Pos 3.5"

FIGURE B Viewing an annotation

Footnotes and endnotes are supported most often through the insert menu, while you can usually access the document properties dialogue box through the file menu. Typically, you can make only limited comments to the document properties.

between the document and your comments file. In addition, you can use the comments file as the basis for a revision plan. Recording your comments in a separate file allows you to focus on the key issues you raised during your review, rather than on individual comments as they appear in a document.

Use the Summary Tool to Check Content and Organization. Summary tools have recently been added to leading word processing programs. Summary tools

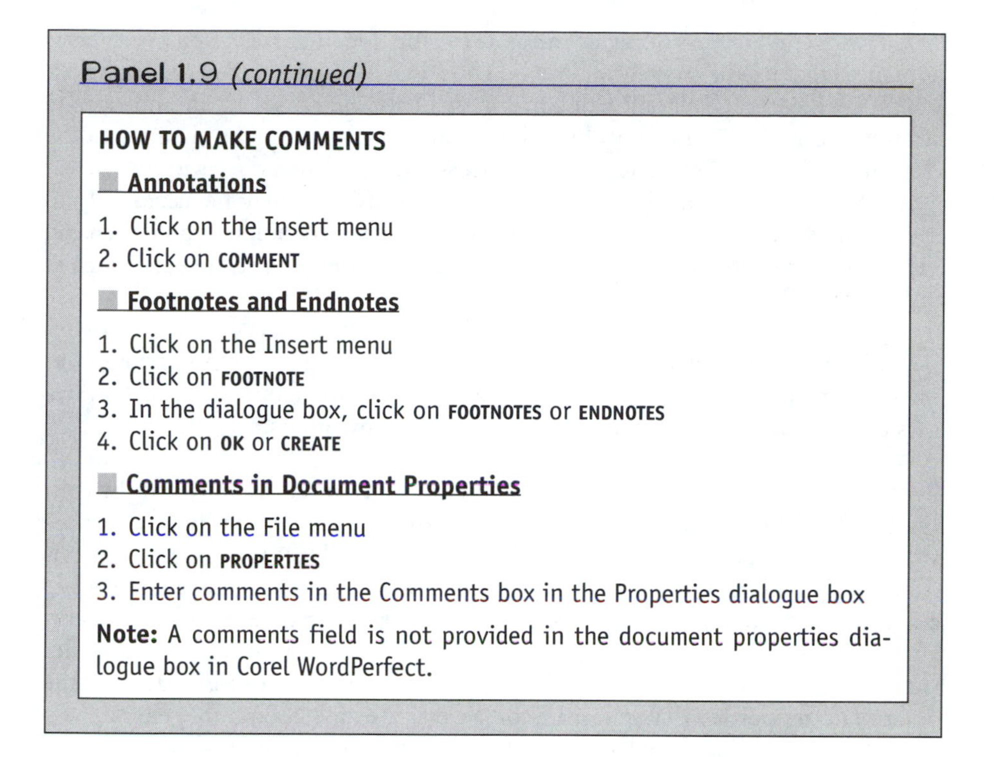
Panel 1.9 (continued)

HOW TO MAKE COMMENTS

Annotations

1. Click on the Insert menu
2. Click on COMMENT

Footnotes and Endnotes

1. Click on the Insert menu
2. Click on FOOTNOTE
3. In the dialogue box, click on FOOTNOTES or ENDNOTES
4. Click on OK or CREATE

Comments in Document Properties

1. Click on the File menu
2. Click on PROPERTIES
3. Enter comments in the Comments box in the Properties dialogue box

Note: A comments field is not provided in the document properties dialogue box in Corel WordPerfect.

are relatively new and may prove useful on shorter documents or on documents that make extensive use of headings. Consider using a summary tool as an additional, but not as a primary strategy to review your document.

Reviewing and Revising Documents Written by Others

Many writers are asked to review or revise documents written by others. Reviewing documents written by others is a common practice in both academic and professional settings. Many writing teachers encourage the practice because they recognize that writers can have difficulty finding problems in their documents. Because writers know what they want to say, they sometimes find it difficult to determine whether they've said it effectively. Asking someone else to review a document can help you discover whether you're making your points effectively.

In professional settings, reviewing documents written by others is also a common practice. Important documents are routinely circulated among members of a writing team, for instance. In many cases, writers in professional settings are asked both to review and revise a document. Managers, for instance, often review and revise documents written by employees they supervise.

The overall process of reviewing and revising documents written by others is similar to that used to review and revise your own documents. You read the document to identify potential problems or inconsistencies, select an appropriate strategy for addressing the problem or inconsistency, and make changes to the document. Two key differences in the process are that you as the reviewer are less familiar with the document than its author and that you may be asked only to review the document. If you are asked only to review a document, you'll focus on identifying problems and inconsistencies and recommending strategies for revising them.

Whether you are asked to review and revise or just to review a document, you'll be able to use many of the strategies that help you review and revise your own documents. In addition to those strategies, we offer four additional strategies for reviewing and revising documents written by others.

Read First—Correct Later. As we noted earlier, it's often useful to begin your review of a document by reading it quickly without making revisions. This provides you with an overall sense of the document and reduces time wasted on revisions that subsequent review may indicate are unnecessary. When you are revising a document written by someone else, avoid making any changes to the text or creating extensive notes about suggested changes. If you must make notes, consider the strategies discussed above for using annotations, footnotes, endnotes, the document properties dialogue box, or a separate document. In general, you should make only brief notes—if any at all—on the first reading of a document.

Highlight Key Passages and Use the Gist and Predict Strategy. When you work with a document that you did not write, you can employ a range of strategies to anticipate where a particular passage will lead you. One of the most useful strategies is called "gist and predict." This strategy asks you to identify the "gist" of a passage—its central idea or crux—and then predict what will come next.

You can mark key passages using the HIGHLIGHT command and record your gist of each passage and your prediction of what will follow in annotations or footnotes. If you find it difficult to predict what comes next, you can return to your annotations for further review.

Create a Document Summary to Identify Key Points. Many reviewers use document summaries to identify the key points and overall structure of a document. This kind of summary stands as a brief paraphrase of the document under review. By viewing the document summary, you can trace the overall line of argument in a document, review its key points, and understand its overall organizational structure. You can write the summary in a separate document, in an annotation or footnote, or, if your word processing program supports it, in the comments section of the Document Properties dialogue box.

Role Play the Audience. A similar strategy is to attempt to role play the audience of a document. As you read the document, ask questions about the document and write your responses in another document or in the comments section of the Document Properties dialogue box.

Reviewing and Revising Documents Collaboratively

Collaborative review of documents is common in academic and professional settings. Writing teachers often ask students to work together in collaborative review groups. Increasingly, teachers are also assigning collaborative writing assignments to groups of students. Students involved in such groups must decide how best to review and revise the document they will submit for a grade.

In professional settings, writers commonly engage in collaborative review and revision. Frequently, the review and revision process is sequential—that is, the document is passed from writer to writer until it is judged (by a manager or the group) to be ready for publication. Less frequently, groups of writers meet to review and revise a document.

Whether you collaboratively review and revise a document one writer at a time or meet as a group, the computer can help you coordinate your efforts. Below, we discuss five strategies that you can use to support collaborative review and revision.

Use Annotations to Support Sequential Review. Annotations not only allow you to read specific comments on a document, but also to identify who made the comments. If you find yourself ready to make a comment on a particular passage or issue, check to see what other reviewers have written. Rather than writing an extensive annotation, you might find it sufficient to note that you agree with the comment made by another reviewer. Moreover, if several reviewers make similar comments on a text, you can be confident that you've identified something that should be addressed during revision.

Use the Document Comparison Tool to Support Sequential Review and Revision. The Document Comparison tool is particularly useful when you are one of several people reviewing and revising a document. If multiple drafts of the document have been saved, or you are able to create multiple drafts using Versions tools, you can use the Document Comparison tool to identify changes made to a document and to determine who made each change.

Use the Computer to Take Notes on Face-to-Face Review and Revision Sessions. If you are reviewing and revising as a group during a face-to-face meeting, ask one person to record ideas and make changes to the document. You can also pass the keyboard from person to person as each reviewer comes up with an idea.

Use Workgroup Software to Review and Revise. If you are working as a group, consider using workgroup software, such as Lotus Notes, to review and revise a document. Workgroup software typically allows each member of the group to make changes to a document, even when they are working at different times or are thousands of miles apart.

Use Chat or a MOO to Role Play the Audience or a Devil's Advocate. Just as you can role play an audience during an individual review of a document written by another writer, you can use role play strategies during a collaborative review. A useful technique is to meet on a Chat channel or MOO to discuss a document. (You can read more about Chat and MOOs in Chapter 4.) You can take on the role of the target audience or appoint members of the group to serve as devil's advocates. Record a transcript of the session for subsequent review and revision.

LOOKING AHEAD

In this chapter, we call your attention to a wide range of strategies for making your computer part of your writing process. Whether you're generating ideas for a writing project, collecting information, planning, drafting, organizing, or reviewing and revising your document, the computer can help you write more efficiently and productively. Because each writer's processes and strategies are so individualized, you're likely to find that some of the strategies we discuss are more compatible to your approach to writing than others. Regardless of the strategies you find most useful, bear this advice in mind as you look ahead to your next writing project: think of your computer as more than simply a glorified typewriter, and you'll save time, write more quickly, and make faster progress in your efforts to improve as a writer.

chapter 2

Editing

Reviewing, revising, and editing your own writing isn't easy. It's often easier to work with another person's writing than your own. The difficulty stems from several sources: from our personal investment in our writing, from the difficulty of seeing our writing through the eyes of others, and from knowing intimately the purpose and content of our drafts.

Writers often become extremely invested in their drafts. Just as we often overlook faults in friends or members of our family, we tend to overlook weaknesses in our drafts. A passage that might strike a reader as wordy or ridden with clichés might strike us as comprehensive or cleverly worded. We've worked hard, after all, to write the draft and it's difficult to tell ourselves that all that hard work hasn't helped us achieve all of our goals for the document.

In addition, it's often difficult for us to put ourselves in the place of our readers. If you're writing for an unfamiliar audience—someone from another country or whose life experiences differ radically from your own—you may find it difficult to predict whether your audience can follow a particular passage or whether the words you've chosen will be hard for them to understand. Even if you're writing for an audience that you know well, it's unlikely that you'll know them as well as you know yourself. The result, unfortunately, can be text that is quite clear to you but difficult for your audience to understand.

A final difficulty with reviewing, revising, and editing our own writing is that we know exactly what we want to say. Because of that knowledge, it's often difficult to determine whether we've provided a clear explanation or sufficient detail to get our points across. A few brief sentences may be enough to remind us of the point we want to make. For our readers, however, much more detail is needed to explain the point.

In the previous chapter, we discussed reviewing and revising text. In this chapter, we extend that discussion to include editing. Editing differs from reviewing and revising in that it focuses on issues at the level of individual paragraphs and sentences. In other words, review and revision focus on larger issues, typically those that concern our overall goals for a document and its various sections, while editing focuses on issues such as the clarity and accuracy of specific passages.

As is the case with review and revision, computers and word processing software offer tools that can help you edit your documents. These tools are particularly useful for checking spelling, checking grammar and style, editing for consistency and variety, and cutting clutter.

EDITING SPELLING

Although the adage "You can't judge a book by its cover" has merit, many readers will judge your documents unworthy if they find spelling and grammar errors. Fortunately, as word processing programs have advanced, they have become more sophisticated for checking spelling errors.

In this section, we provide an overview of the Spell Checking tool and explain how to use it. Spelling checkers can help you find and correct spelling errors. They catch mistakes you may have missed when reading your document on the screen. Why? Research has shown that reading on the screen is slower and more difficult than reading printed text. Research also suggests writers don't read their text as well as that of other writers because they know what they want to say.

Spelling checkers are particularly valuable because they allow you to focus on your ideas, your audience, and your purpose as you write—rather than on lower-level concerns with spelling. When you draft documents, turn off your internal editor—that haunting voice that says everything must be correct as soon as you see it on the screen. Keep the words flowing. Don't worry about spelling errors. You can correct them later with the spelling checker. Using a spelling checker in this way enables many writers to generate text more quickly and efficiently.

When it's time to use a spelling checker, don't assume that it will catch every problem. Because of the difficulties associated with reading text on the screen, reread your text on paper. Even highly skilled writers and editors find errors in printed copies of a document that they missed when reading text on a screen.

Spelling checkers can speed your editing and minimize spelling errors in your final documents—if you understand their strengths and weaknesses. Spelling checkers are a boon to writers because they make automatic corrections, make immediate checks, and provide an "electronic editor." However, spelling checkers have several limitations. Spelling checkers:

- May flag correctly spelled words as being wrong
- Fail to distinguish between homonyms (words that sound alike but are spelled differently)

- Have limited dictionaries
- May use different spellings for the same word
- May slow the writing process

Consider each point in more detail. When writing quickly, you may type the wrong word, but spell it correctly. If so, the spelling checker program reads the correctly spelled wrong word and does not flag it. Since you typed the wrong word, but spelled it correctly, the spelling checker does not flag the word as being misspelled.

Homonyms, words that sound alike and may have the same spelling but that have different meanings, present a problem for spelling checkers. If you type *two* when you mean *to* or *too*, a spelling checker will not tell you that you've used the wrong word. That said, grammar and style checkers sometimes flag words with common homonyms.

Dictionaries in word processing programs, like most printed dictionaries, do not include all words. General dictionaries may not contain specialized words used in technical, scientific, or other specialized fields and organizations. Nor do they contain formal names and titles used in many organizations. For example, you will not find the word *sunspace*, a greenhouse or sunroom attached to the south side of a house, in many collegiate dictionaries. Furthermore, printed dictionaries may contain one or more acceptable spellings for the same word, while software dictionaries may contain only one acceptable spelling. Fortunately, spelling checkers recognize words not in their dictionaries and allow you to add new words to the dictionary.

Keep in mind that different dictionaries may spell some words differently. If you are using a style guide that specifies use of a particular dictionary, you may need to check the spelling against the dictionary. If you're a student writing term papers or reports for classes, ask your instructor which style guide and dictionary to follow.

tip 2.1 ■ Using the Spelling Checker

To capitalize on the strengths of spelling checkers:

1. Run the spelling checker on an entire document once you've finished a draft.
2. Don't assume that the spelling checker is always correct.
3. Run the spelling checker twice. At times, you may introduce another spelling error when running the program.
4. Print a copy of the document.
5. Edit the printed copy of the document.

Using Automatic Flagging Spelling Checkers

Recent word processing packages can flag possible misspelled words as you write. If you're using a word processing program with this feature, you can wait until you finish drafting the document or you can make the corrections as you work. When you are writing and the spelling checker signals a potential spelling error (Figure 2.1):

1. Click on the right mouse button
2. Review the menu of the possible correct spellings
3. Click on the correct spelling

The spelling checker will replace the misspelled word with the correctly spelled word.

If you find that stopping to correct spelling errors disrupts your composition, you can turn off the feature or ignore the flagged spelling errors and correct them later. If you choose to wait until after you have finished drafting your document:

1. Move the cursor to the top of your document
2. Save your document to the hard drive
3. Use SAVE AS to make a backup on a removable disk such as a floppy diskette

To correct misspelled words:

1. Scroll down looking for flagged errors
2. Stop on each flagged error
3. Click on the right mouse button
4. Scan the list for possible correct words
5. Highlight the correct word, if listed
6. Click on it

The software should replace the incorrectly spelled word with the correct word.

If the system does not provide a correct spelling of the word, type in the correct spelling. If you don't know the correct spelling, check an acceptable printed dictionary, such as *Webster's Tenth New Collegiate Dictionary*. Check the style manual that you're using to see what dictionary the style manual editors recommend. If you're writing about a scientific or technical subject, you may need to check specialized dictionaries (see Appendix D for a list of accepted dictionaries).

Turning Off Automatic Spelling Checkers. If you find that the automatic flagging of spelling errors disrupts your writing, turn off the spelling checker. First, use HELP from the menu or button bar, then search for the Options, Preferences, or User Setup dialogue boxes in your word processing program. Review the online directions to learn which menu will allow you to access the dialogue box. Keep in mind that some word processing programs use different names and locations for accessing default settings.

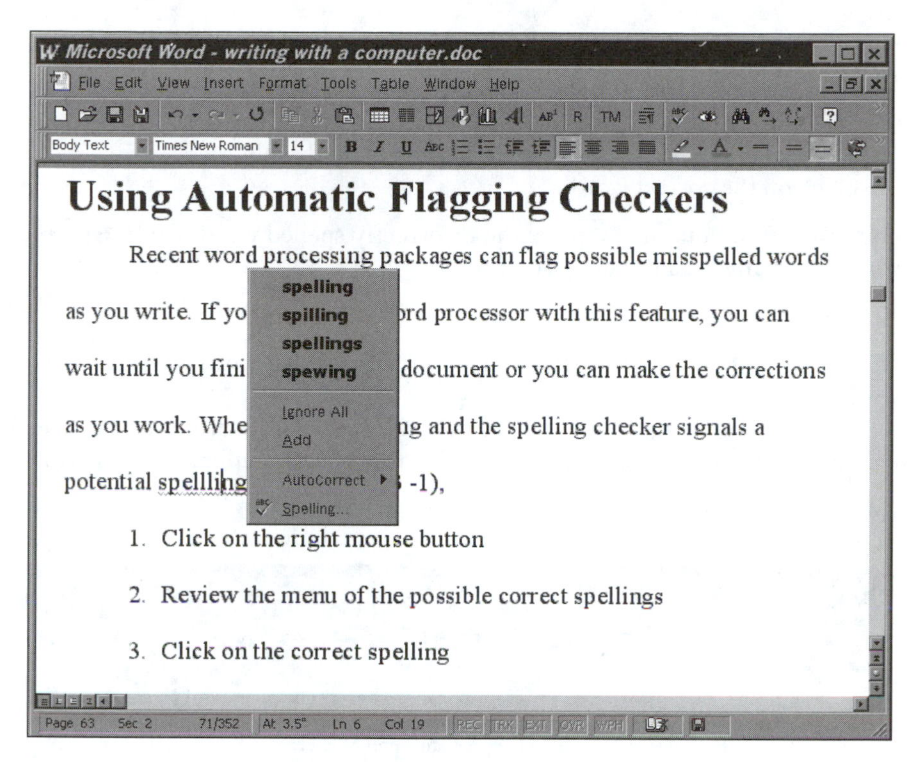

FIGURE 2.1 Automatic flagging of a misspelled word

Using Command Driven Spelling Checkers

If you're using a spelling checker in an older word processing program or you have turned off the automatic flagging feature, you can invoke the spelling checker and review your draft. To use the spelling checker:

1. Scroll to the top of the document
2. Use SAVE to save your document to your hard drive
3. Use SAVE AS to save the document to a backup or removable disk
4. Use SAVE AS to save the document back to the hard drive

Then run the spelling checker using the menu or by using the spelling checker icon on the button bar. To start the spelling checker via the menu:

1. Click on the Edit or Tools menu
2. Click on SPELLING

The spelling checker will start (Figure 2.2) searching the document. When it stops on a misspelled word:

1. Scan the suggestions box for the correct spelling of the word
2. Place your cursor on the correct word if it is not highlighted
3. Click on the correct word

The software should then replace the incorrectly spelled word with the correct word. Alternatively, another option is to:

4. Click on the **REPLACE** button

The software should then replace the incorrectly spelled word with the correct word.

The spelling checker dialogue box usually has additional buttons to speed making spelling corrections. You can add words to the dictionary or the Auto-

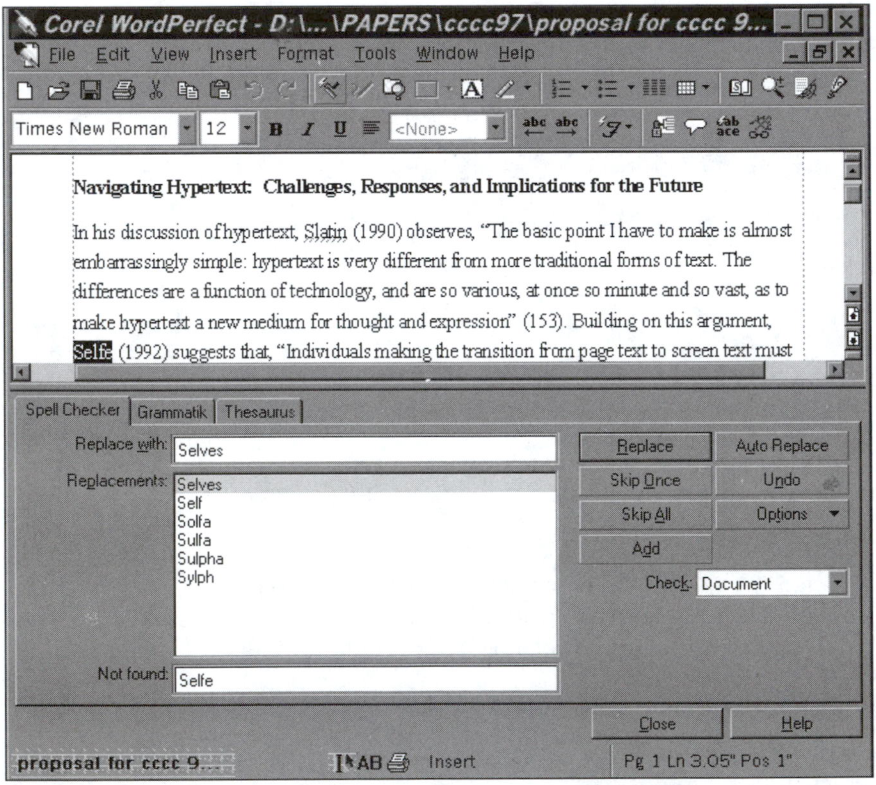

FIGURE 2.2 Spelling dialogue box

correct tool. You can ignore specific words by clicking on the **SKIP** or **IGNORE** buttons, or you can have the program change all words in the document that you have misspelled the same way.

Explore the spelling checker the first time that you use it and review the online help available by clicking on **HELP** and searching for terms such as spelling or spelling checker. Although keystrokes may vary among word processing programs, the end result is the same—the spelling checker helps you catch errors and make changes.

EDITING FOR GRAMMAR AND STYLE

Grammar and style checking tools provide a quick way to review your draft for basic grammar, mechanical, and style problems. They work much like the spelling checking tool and, like spelling checkers, grammar and style checkers have both strengths and weaknesses. Like spelling checkers, grammar and style checkers are not always accurate. Consider these five major reasons:

1. Because American English draws words from many different languages, you'll find hundreds of exceptions to the rules. The mix of words from different languages gives English its richness, and creates a wide range of possibilities for each sentence you write.

2. Grammar and style checkers cannot identify all the permutations of language constructions that might arise in any one text. Although major corporations, such as IBM and AT&T, worked for years on sophisticated programs designed to analyze style, the programs could not distinguish all of the nuances used by writers.

3. Acceptable usage varies from discipline to discipline and publication to publication. Hundreds of style manuals have been compiled and many businesses and organizations have their own style manuals (see Appendix D, Reference Works and Style Guides). That said, grammar and style checkers follow commonly accepted rules.

4. Different writers, writing teachers, and editors interpret the rules differently. What one writer, teacher, or editor says is right, another may say is wrong. In the long run, consistency emerges as the most important rule. Major style manuals include *The MLA Style Manual* (commonly followed in the humanities), the *Publication Manual of the American Psychological Association* (commonly used in the social sciences), the *Associated Press Stylebook* (commonly used in the newspaper and magazine industries), the *Council of Biology Editors Style Manual* (commonly used in the biological sciences), and *The Chicago Manual of Style* (commonly used in the publishing industry). When editing your drafts, select the style manual appropriate to the publication or audience for which you're writing and then follow that style manual consistently.

5. English evolves over time and experts debate some points. Over the decades, hardened rules give way as usage dictates changes. For example, many major style manuals suggest using *that* to introduce restrictive clauses and *which* to introduce nonrestrictive clauses. Read carefully many popular and semi-popular publications, however, and you'll see that *which* frequently is used in place of *that*.

Although they aren't perfect, grammar and style checkers also have at least four advantages. First, grammar and style checkers make use of hundreds of rules. When you run a grammar and style checker, the program checks your draft against all of the rules. If your text does not conform to the rules, the program flags the text passage and suggests a change. Second, grammar and style checkers do not suffer the fatigue that you may suffer after working on your documents for several hours. Third, unlike writers, grammar and style checkers keep on task and are not distracted by outside thoughts. Fourth, grammar checkers provide a second set of "eyes" to review your drafts. For these reasons, use grammar and style checkers to review your drafts and to catch errors that may have slipped into them.

If you use a word processing program that automatically flags grammatical and stylistic errors as you write, you can treat it in the same way that you deal with automatically flagging spelling errors. If the automatic checker disrupts your writing, turn it off, then run it after you have a near-final draft. To learn how to turn off your automatic grammar and style checker, click on **HELP** and search for information on how to configure the grammar and style checker on your word processing program.

To use grammar and style checkers effectively, you need a solid understanding of grammar and style rules and how to apply them. When a grammar or style checker flags a text passage, you need to review the passage and decide whether or not it violates a grammar or style rule. If you're unsure, click on the checker's **EXPLAIN** or **RULES** button, and it will display the rule. If you are still unsure of the suggested change, consult a grammar and style handbook, or a collegiate dictionary (see Appendix D for a list of style manuals, dictionaries, and grammar handbooks).

tip 2.2 ■ Using the Grammar and Style Checker

To capitalize on the strengths of grammar and style checkers:

1. Run the grammar and style checker on the entire document once you've finished a draft.
2. Don't assume that the grammar and style checker is always correct.
3. Run the grammar and style checker twice. At times, you may introduce another grammatical or stylistic error when running the program.
4. Print a copy of the document.

Keep in mind that grammar and style checkers frequently flag passages that do not violate grammar and style guidelines. If you accept suggested changes without reviewing the passages, you may introduce additional errors into your document.

Using Automatic Grammar and Style Checkers

As with spelling checkers, recent word processing programs include automatic grammar flagging programs. As you write, the programs identify potential problems by highlighting or placing colored wavy lines under the passage that may violate the grammar rules programmed into the word processing program. You can ignore the flagged constructions until you've finished writing your draft and then return and review the flagged construction, or you can review the potential errors as you write.

Keep in mind that all flagged constructions are not necessarily wrong (Figure 2.3). You'll need to review each flagged passage, then decide whether to

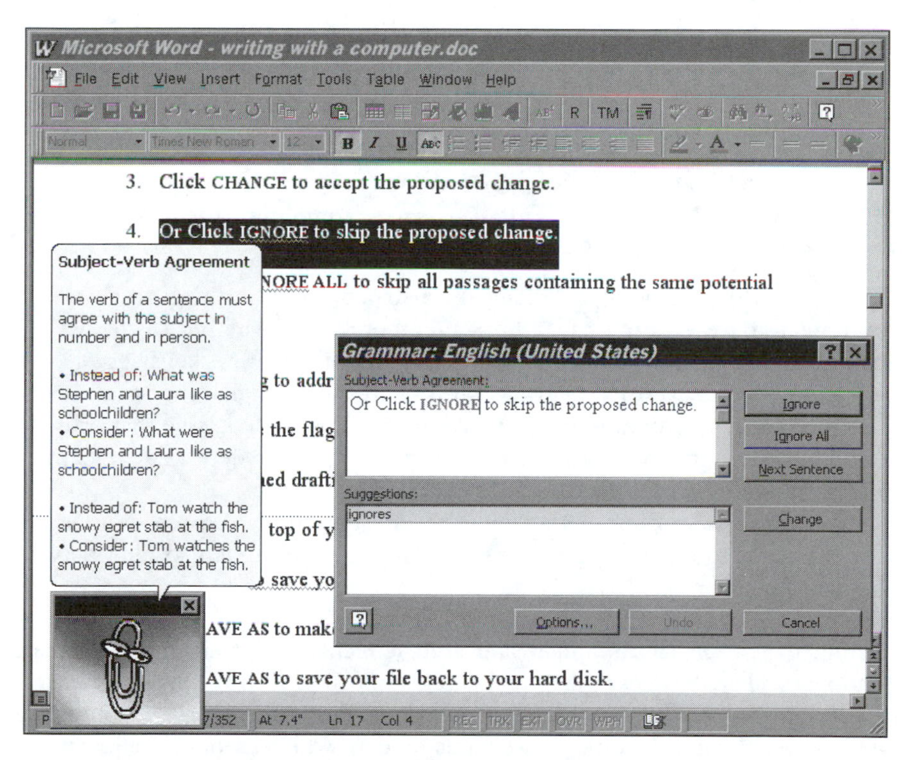

FIGURE 2.3 Grammar and style dialogue box

change it or not. If you agree that the passage should be improved, you'll need to revise it. To use the automatic flagging programs:

1. Click on the right mouse button
2. Review the suggested change
3. Click CHANGE to accept the proposed change
4. Or click IGNORE to skip the proposed change
5. Or click IGNORE ALL to skip all passages containing the same potential problem

If you find that stopping to address grammar and style issues disrupts your writing, turn off the feature or ignore the flagged passages and deal with them later. If you decide to wait until you've finished drafting your document:

1. Scroll to the top of your document
2. Use SAVE to save your document on the hard drive
3. Use SAVE AS to make a backup copy on a removable or backup disk
4. Use SAVE AS to save your file back to your hard drive
5. Start the grammar and style checker by clicking on an icon on the button bar or by using the menu

To correct potential grammatical and stylistic problems:

1. Click on the right mouse button
2. Review the suggested change
3. Click CHANGE to accept the proposed change
4. Or click IGNORE to skip the proposed change
5. Or click IGNORE ALL to skip all passages containing the same potential problem

Again, keep in mind that all flagged constructions are not necessarily wrong. You'll need to review each flagged passage, then decide whether to change it or not. If you do not understand why the grammar and style checker flagged a passage, click on the EXPLAIN or RULES buttons in the grammar and style checker dialogue box. The program will display the grammar or style rule that provided the basis for flagging the passage. Review the rule and then determine if it applies to your passage.

Recent versions of automatic grammar and style checkers catch irregularities in text such as too many spaces between sentences, too much spacing between punctuation, and related problems.

Turning Off Automatic Grammar and Style Checkers. Again, if an automatic grammar and style checker disrupts your composing, turn it off. Use the HELP from the menu bar, then search for the Options, Preferences, or User Setup dialogue boxes. Review the online directions to learn which menu will allow you to access the respective function. Keep in mind that some word processing programs use different names and locations for accessing default settings.

Using Command Driven Grammar and Style Checkers

If you're using a version of a word processing program that does not have automatic flagging of grammar errors or you have turned off the automatic flagging feature, you can invoke the grammar checker to review your draft. To use the grammar checker:

1. Scroll to the top of the document
2. SAVE the document to the hard drive
3. Use SAVE AS to save the document to a removable or backup disk
4. Use SAVE AS to save the document to the hard drive again

Then run the grammar checker using the menu bar or by using the grammar and style checking icon on the button bar. To start the grammar and style checker via the menu:

1. Click on the Edit or Tools menu
2. Click on GRAMMAR

The grammar and style checker should start. When it stops on a passage:

1. Review the passage and the flagged item
2. Review the suggested changes
3. Click on the suggested change, if you accept the recommendation
4. Click on the RULE, HELP, or EXPLAIN button if you don't understand why it's flagged
5. Review the explanation
6. Make the changes if you think the suggestions apply to your passage

At times you may need to return to your draft and rewrite the passage to eliminate the problem. To do so:

1. Click outside of the grammar and style checker window
2. Revise the passage
3. Return to the grammar and style checker and resume checking
4. Continue evaluating the suggested changes

A Cautionary Note

As you use Spell Checking and Grammar and Style Checking tools, remember that flagged words or passages may not be wrong. Current versions of these tools cannot, at this time, analyze all the complexities of spelling, grammar, and style.

If you use them with a clear understanding of their limitations, however, you'll find that they can help you polish your documents and improve your writing.

tip 2.3 ■ Customizing the Grammar and Style Checker

Whenever you use a grammar and style checker, check each flagged item to decide whether or not you want to accepted the recommended change. Keep in mind that the recommended changes are based on a limited number of editorial style rules. For example, we used an informal style when drafting the manuscript for this book—in particular, we used personal pronouns and contractions to make the text more conversational. When we ran the grammar checker on our introduction, it stopped on the third paragraph and flagged the contraction, *they're.*

The grammar and style checker told us, when we clicked on the explanation button, that contractions were inappropriate in formal prose (Figure 2.4). Since we were concerned with creating a more conversational tone, we turned the contractions rule off.

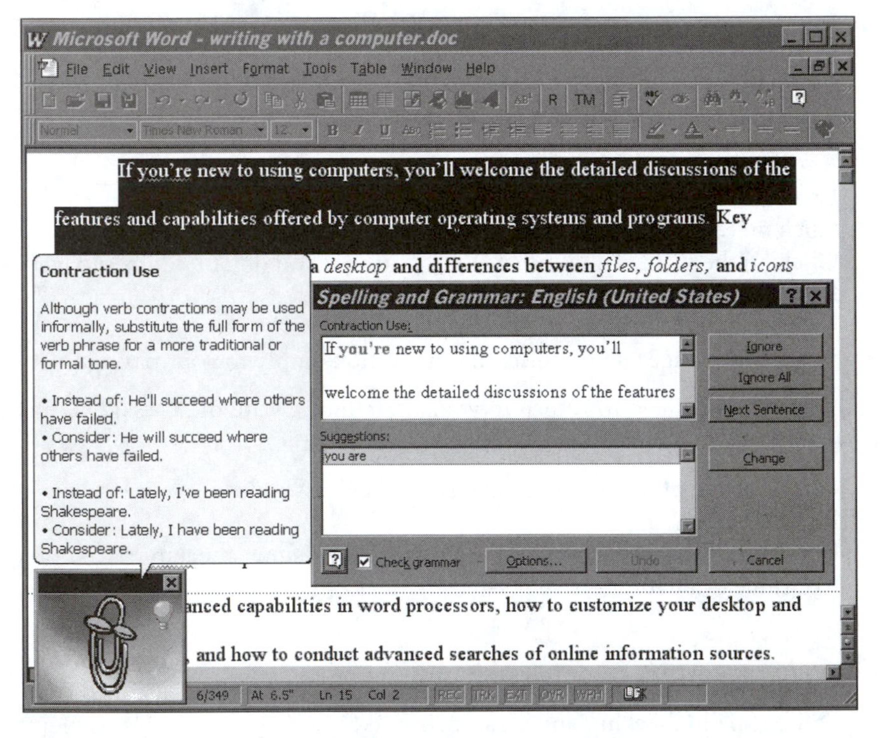

FIGURE 2.4 Grammar and style advice on contractions

EDITING FOR CONSISTENCY AND VARIETY

Experienced writers edit for spelling, grammar, and style because they know that the small things make a difference. A misspelled word, a grammatical error, or an awkward phrase can call a reader's attention away from the message you're trying to convey in a document. It isn't that a few misspellings, punctuation errors, or inelegant sentences will make it impossible for a reader to understand what you're trying to say, but that these kinds of errors can undermine your credibility.

For the same reasons, experienced writers edit their documents for consistency. They carefully check their prose to determine, among other things, whether they've used facts and figures accurately, described key ideas in the same way, and spelled names and titles consistently. You'll certainly lose credibility, for instance, if your readers notice that you've referred to Bill Clinton as President Clinton in one sentence and as President Clinten in the next. They're likely to be even more concerned if you argue that we need to increase spending on education by 8.7% in one part of your document and 87% in another.

Readers are also likely to grow annoyed with a document if, time and again, they find the same phrases or words used to describe something. You can read the phrase *innovative technique* only so many times before you start to ask why the writer couldn't be a bit more innovative with adjectives. Readers, like any audience, enjoy variety. For this reason, experienced writers also edit their writing for variety.

Editing for consistency and variety are areas in which computers have significantly eased the burden for writers. You can use a range of computer-based techniques to edit for consistency and variety, among them the **FIND** command, the **REPLACE** command, and the **VIEW OUTLINE** command.

Editing for Consistency with the FIND Command. It's easy to edit for consistency when you're writing a two-paragraph memo, but it's quite another thing when you're writing a 35-page report. You can use the **FIND** command, however, to quickly scan your document for key words and phrases. In some cases, you'll simply want to make certain that you've spelled someone's name correctly. In other cases, you'll want to make sure that you've used the correct budget figures throughout a document. In more complex cases, you'll want to determine whether you've defined a key idea in the same way throughout your text.

For instance, you might write a report in which you argue that a certain set of climatic changes will lead to lower grain production in Iowa, Illinois, Wisconsin, and Minnesota. Such an argument is likely to rely on numerous figures and on a carefully constructed causal model. To edit for consistency, you can pick out key words or phrases that are typically used in conjunction with your figures—a phrase such as *grain production* would be one example. Using the **FIND** command, you can locate each instance of the phrase *grain production* and check any figures that appear in the surrounding text.

Editing for consistency is aided greatly by the FIND command. Using the FIND command, you can quickly locate key words and phrases, check the spelling of names, titles, and terms, and make certain that you've used facts and figures consistently.

Editing for Variety Using the FIND Command. You can use the same techniques to edit for variety. While editing for consistency often deals with issues of accuracy, editing for variety on the other hand is almost always a stylistic concern. Experienced writers edit for variety to keep their readers interested in what they have to say. Readers can quickly grow weary of repetitive prose such as, "The brilliant young scientist, well known for his brilliant discussions of the most esoteric topics and equally well known for his brilliant insights into the cosmos, elaborated this with his characteristic brilliance."

You don't need to be brilliant, of course, to edit for variety—although it helps. Simply read your document with the goal of finding phrases, modifiers (adjectives and adverbs), and verbs that are used repeatedly. There's nothing wrong with repeating yourself a few times, particularly in longer documents, but you should avoid reusing the same words and phrases when alternatives can be found.

Once you notice that you've used a word or phrase too often, use the FIND command to see how often you've actually used it. Use the FIND NEXT command to move quickly through your document. You'll soon know whether you've actually overused a word or phrase.

Editing for Consistency and Variety with the REPLACE Command. Finding inconsistencies and overused words and phrases is a great start, but you'll need to follow through by changing your text. Often, the changes are few and simple. Occasionally, however, you'll find that you must make dozens, or even hundreds of changes to a document. Rather than making these changes one at a time, you can use the REPLACE command to make wholesale or selective changes.

Use REPLACE to correct inconsistencies or to correct inaccuracies. If you've used the wrong number throughout a document, use replace to change it to the correct number. You can use the REPLACE ALL option in the REPLACE dialogue box to change all instances of a word, phrase, or number in a text. But be careful: replacing all instances of a word, phrase, or number can be dangerous. Your computer won't be able to distinguish between a word that should be changed and one that should not. For instance, you might decide to change all instances of *MS* (as in *MS Word*) to *Microsoft*. If you don't select specific options in the REPLACE dialogue box (Figure 2.5), you'll find yourself changing words like *seems* and *themselves* to *seeMicrosoft* and *theMicrosoftelves*.

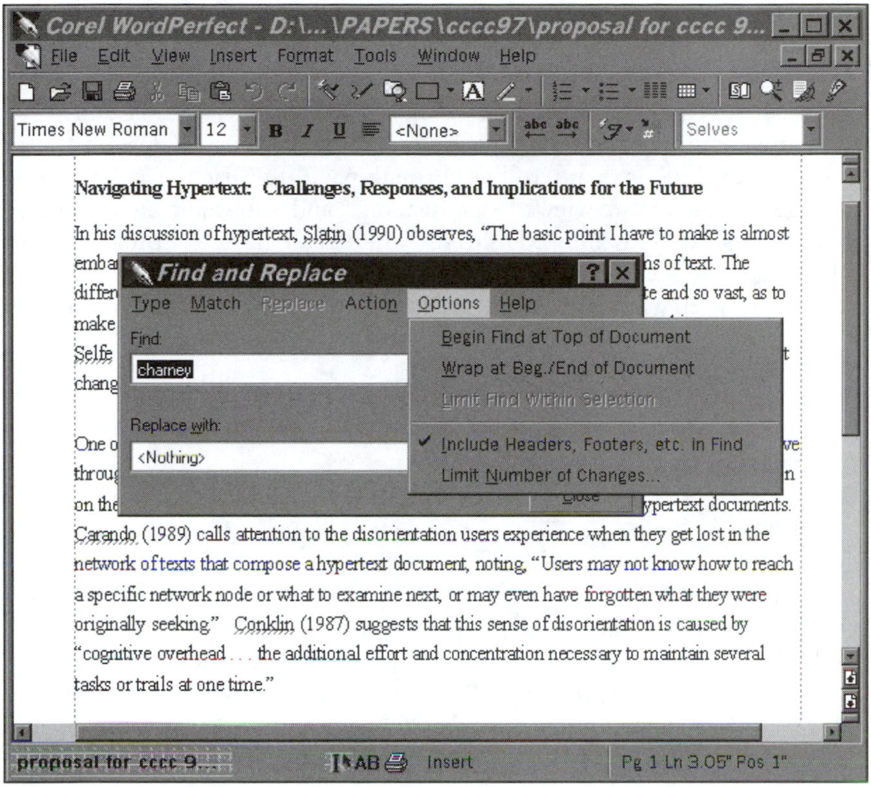

FIGURE 2.5 FIND AND REPLACE options

You can avoid these intriguing transformations by specifying options in the Replace dialogue box such as **MATCH CASE** and **FIND WHOLE WORDS** only. In most word processing programs, **REPLACE** allows you to specify other options as well, including formatting (useful, for instance, if you'd like to change underlined words to italicized words) and searching only selected text.

You can also use **REPLACE** to make selective changes. If you are editing for variety, you can reduce your use of specific words and phrases by using the **FIND NEXT** and **REPLACE** buttons in the Replace dialogue box. For instance, there's nothing wrong with the word *brilliant*, so you wouldn't want to completely eliminate it from your document. Instead, use the **FIND NEXT** and **REPLACE** buttons to selectively change the word. If you've used *brilliant* 25 times in a 25 page document, you might want to substitute it with three words: *insightful*, *intelligent*, and *smart*. Make three passes through your document, selectively replacing some instances of *brilliant* with *insightful*, *intelligent*, or *smart*. You'll end up with more variety in your document, something most readers appreciate.

Editing for Consistency and Variety with Outlining Tools. Outlining tools allow you to view headings and text in a document selectively (see Chapter 1). To use these tools effectively, you need to format your document using **STYLES**, which we discuss in the next chapter.

We used outlining tools to edit this book for consistency and variety in our use of headings. Rather than scrolling through page after page of a several hundred page document, we viewed our headings and subheadings without the accompanying text. This allowed us to see whether we had constructed our section headings consistently and whether we had sufficient variety in our headings. We could easily tell whether we were overusing particular verbs and whether our verbs were in the same forms. We could also tell whether we had organized sections consistently.

Cutting Clutter

If you're like most writers, you can trim your first drafts by 10%, 20%, or 30%—perhaps even more. Cutting the clutter—reducing wordiness and choosing more appropriate words—often makes your writing clearer and more forceful, qualities that usually lead to more effective communication.

The key to cutting clutter focuses on identifying problems and recasting your text to eliminate the problems. By using word processing commands such as **FIND**, you can quickly identify potential problems and recast passages. Specifically, you can use your word processing program to help you:

- Recast passive voice constructions
- Release trapped verbs
- Remove "There are . . ." and similar constructions
- Edit excess prepositional phrases
- Edit excess clauses
- Use shorter words
- Remove unnecessary qualifiers
- Eliminate wordiness
- Eliminate vague, general words
- Avoid unnecessary jargon

The primary reason to use the computer to identify wordy passages is to help you learn to recognize wordy and unclear constructions in your writing. While the following discussion identifies 10 wordy constructions, you may find it too tedious and time consuming to use **FIND** command to check your drafts for each construction. If so, begin by using your word processing program to search for passive voice constructions and trapped verbs. As time permits, search for the other wordy constructions. Over time, you'll learn to recognize problems in your drafts and recast problematic passages without using your word processing program. Below, we consider each technique in more detail.

Recast Passive Voice Constructions. When revising, look for passive voice constructions. Passive constructions tell you that something was done, but do not tell you who did it. The following sentences use passive voice:

The report was completed.

The project plan was begun by the committee.

In contrast, the following sentences use active voice to tell who carried out an action:

The committee completed the report.

The committee began planning the project.

Communication research suggests that readers understand active voice more easily than passive voice. Because of this, documents that make extensive use of passive voice can be comparatively difficult for readers to understand. This doesn't mean that you should always avoid passive voice, but it does suggest that you should be cautious about using it. Appropriate uses of passive voice include situations in which the actor is unknown or in which you do not want to report who is carrying out the action. You can also use the passive voice when you want to place the focus of attention on someone or something other than the actor. For example:

The child was struck by a car while crossing the street.

In such constructions, the subject is more important than the actor. Typically, you can use passive constructions:

- To focus attention on the subject
- To stress the points at the end of a sentence
- To provide variety

Consider further such constructions as

Active Voice:

John dropped the computer.

The sentence answers the questions, "What happened?" or "Who is responsible?"

Passive Voice:

The computer was dropped.

The computer was dropped by John.

Passive constructions answer such questions as, "Why didn't . . . ?" or "What seems to be the problem?"

Passive constructions often leave readers wondering who did what and usually require longer sentences. In contrast, active voice constructions use stronger verbs and often, but not always, produce shorter, more vigorous sentences, and make clear who did what.

To recognize passive voice constructions, ask:

- Does the sentence use *to be* verb forms—*is, was, were, have, has*?
- Does the sentence have a prepositional phrase beginning with *by*?
- Does the sentence have an implied *by* type of prepositional phrase?

You can use two word processing tools to identify passive constructions—the **FIND** command or the Grammar and Style Checking tool. To use the **FIND** command to locate possible passive constructions, search for the following words:

is	was	were
have	has	had

Or you can use the Grammar and Style Checking tool in your word processing program. This tool won't tell you whether it's appropriate or inappropriate to use passive voice in a given sentence, nor is it always accurate in its identification of sentences containing passive voice, but it can call your attention to sentences that contain constructions typical of passive voice. In the second section of this chapter, we discuss the Grammar and Style Checking tool in detail.

Whether you are using the **FIND** command or the Grammar and Style Checking tool, examine sentences flagged by your word processing program to determine whether or not you should convert the sentence to active voice. Ask:

- Who carried out the action?
- Was it a person?
- Was it an organization?
- Was it an animal?
- Was it a plant?
- Was it an object?

Then determine whether or not you need to convert the sentence to active voice. If you want to make clear who did what, then use the active voice. Keep in mind that active constructions are usually short and may make your text more understandable. Use passive constructions when the item being acted upon is more important than the actor, when the actor or actors are unknown, or when you want to avoid placing responsibility or blame on a person or organization.

Warning. Do not use the automatic **FIND AND REPLACE ALL** *command. Use* **FIND** *and then examine each construction.* While word processing programs may, someday, analyze the logic or meaning of your sentences, today's word processing programs do not have that capability.

Release Trapped Verbs. Use the **FIND** command to locate words ending in

-tion	-ence
-al	-ment
-ance	-ure

Words ending in such constructions are called nominalizations, and they frequently have a verb form. If appropriate, recast the sentences using the verb.

Draft:

Denver, Colorado, is conducting an experiment with a clean-needle-replacement program to control AIDS among drug addicts.

Recast:

Denver, Colorado, is experimenting with a clean-needle-replacement program to control AIDS among drug addicts.

Draft:

The committee is in the fifth hour of its discussion on the rules.

Recast:

The committee has discussed the rule changes for five hours.

Remove "There Are . . ." and Similar Constructions. Use the **FIND** command to locate sentences beginning with

There are . . .
There is . . .
These are (have) . . .
It has been reported that . . .

Although such constructions are not technically wrong, many use passive voice and thus are candidates for revision. An occasional "There are . . ." construction does not present a problem, but inexperienced writers often repeat such constructions too frequently. If you're striving to improve your document, consider tightening the text by reversing the order of the sentence or deleting unnecessary introductions.

Consider each "*There are* . . ." construction and ask whether or not you can reorder the sentence to make your point. Consider the following example:

Draft:

There are five dogs in the pen.

Recast:

Five dogs are in the pen.

Or,

The pen holds five dogs.

First determine whether the dogs are more important than the pen, and then decide which construction best carries your message.

If the introductory construction adds no meaning to your sentence, delete it. Consider the following example:

Draft:

It has been long known that the West has a water problem.

Recast:

The West has a water problem.

If your "There are . . ." constructions use passive voice, consider changing them to active voice.

Edit Prepositional Phrases. Using too many prepositional phrases makes sentences long and more difficult to follow. For example,

The team of the Division of Corrections of the Department of Human Services makes recommendations of a controversial nature.

To find prepositional phrases, use the **FIND** command to locate prepositions:

at	except	of	through
as	for	on	to
but	from	over	upon
by	in	past	with
during	into	since	within

Edit prepositional phrases by eliminating prepositional phrases, recasting prepositional phrases as adjectives or adverbs, or converting them to possessives.

Consider eliminating prepositional phrases. Look closely at prepositional phrases ending sentences. In many cases you can eliminate them without changing the sentence meaning:

Draft:

Whenever possible, edit prepositional phrases in sentences.

Recast:

Whenever possible, edit prepositional phrases.

Since the prepositional phrase *in sentences* adds no meaning, you should think about dropping it. Once you have established that you are discussing a specific topic, and your document centers on that topic, do you need to keep referring to it? If you think your readers might find your discussion confusing, retain the prepositional phrase.

If you can't eliminate prepositional phrases, convert them to adjectives or adverbs. Such changes provide more concise sentences:

Draft:

The cat of a yellow color pounced on the robin.

Recast:

The yellow cat pounced on the robin.

By converting the prepositional phrase, *of a yellow color*, to an adjective, *yellow*, you shorten the sentence:

Draft:

The mountain lion walked in a rapid manner toward the poodle.

Recast:

The mountain lion walked rapidly toward the poodle.

In the same way, converting the prepositional phrase, *in a rapid manner*, to an adverb, *rapidly*, shortens the sentence.

If you can't delete or recast the prepositional phrases as an adjective or adverb, consider converting the prepositional phrase to a possessive:

Draft:

The tail of the mountain lion was only 6 inches long.

Recast:

The mountain lion's tail was only 6 inches long.

Draft:

The squeal of the poodle . . .

Recast:

The poodle's squeal . . .

In both examples, converting to the possessive eliminates unnecessary prepositional phrases.

Edit Clauses. Putting too much information in sentences with subordinate clauses often burdens the reader with excess verbiage:

The summer camp that was to have begun June 22 was canceled by the youth committee.

Relative pronouns often signal subordinate clauses:

that whom who which what

Edit clauses by eliminating them, converting them to appositives, replacing with adjectives, and replacing them with prepositional phrases. Be cautious, however. Replace a clause with a prepositional phrase only when you must retain an idea to ensure that you have not confused your readers.

To locate clauses in your drafts, use the **FIND** command and then consider what strategy you can use to edit them. In some cases you'll need a clause to make your point. In other cases, revising clauses will improve your writing.

When you find a clause, ask, "Do I really need the clause to make my point? Can I eliminate it? Is it relevant to the text?" Consider the following examples:

Draft:

Club president John Smith, who is a nice guy, spoke at the meeting.

Recast:

Club president John Smith spoke at the meeting.

If the fact that John is a nice guy is irrelevant to the sentence and your document, then eliminate the clause.

You can often shorten a sentence by converting the clause into an apposition —a short description following the word it modifies. Consider the following example:

Draft:

Ed Carpenter, who is the accountant, will review the books.

Recast:

Ed Carpenter, the accountant, will review the books.

The clause isn't needed to describe Ed Carpenter. Whenever you locate a clause, ask whether you can describe the noun by using an apposition.

You can convert some clauses to adjectives. Consider the following example:

Draft:

Mountain lions that are hungry seek easy meals.

Recast:

Hungry mountain lions seek easy meals.

Whenever you locate a clause, ask whether you can convert it into an adjective. In many cases, doing so will help you make your point more succinctly.

If you can't eliminate a clause, create an apposition or adjective, and try to convert the clause to a prepositional phrase. Consider the following example:

Draft:

Prune branches that have no new shoots . . .

Recast:

Prune branches without new shoots . . .

Once you have replaced the clause with a prepositional phrase, ask yourself whether or not you could eliminate the prepositional phrase. In many cases, you can.

Use Shorter Words. When editing, replace long, complex words with short, easy-to-understand words. But don't end up with a "See Spot run. See Jane chase Spot." style. Nor should you write down to your readers. Instead, speak plainly and clearly:

Draft:

I will be in communication with you on Friday.

Recast:

I will talk with you on Friday.

Draft:

We must prioritize the issues facing our community.

Recast:

We must rank the issues facing our community.

Remove Unnecessary Qualifiers. Use the FIND command to locate unnecessary qualifiers such as

sort of	somewhat	soon	several
a little bit	really	lots	kind of
very	a bit	many	really

When you locate them, ask yourself whether they add meaning to your writing. If not, edit them by replacing them with specific words or eliminating unnecessary qualifiers.

Whenever you locate an unnecessary qualifier, replace it with a specific word. Consider the following example:

Draft:

We will soon complete the report.

Recast:

We will complete the report by 3 p.m.

Alternately, consider whether you can delete the unnecessary qualifier. Keep in mind that *very* doesn't mean *very much:*

Draft:

The writer reported that he sort of had a problem with the laser printer.

Recast:

The writer reported a problem with the laser printer.

The writer reported the laser printer will not print.

Eliminate Wordiness. When reading your drafts, question every word, phrase, clause, or paragraph. Ask whether you can make the point in fewer words. Consider the following pairings:

Wordy	*Succinct*
At this point in time	now
In order to	to
Preliminary planning	planning
by means of	by

Consider the following examples:

Draft:

In order to expedite your proposal.

Recast:

To expedite your proposal.

Draft:

At this point in time we will expand our office staff.

Recast:

We will expand our office staff.

Draft:

Printing our publication will be completed by means of computers.

Recast:

Computers will help us print our publication.

Eliminate Vague, General Words. To make your message clear, use specific, concrete terms. Vague, general terms leave readers uncertain about your meaning. Compare the following pairings:

Vague, General Terms	*Specific, Concrete Terms*
In a few days	two days
Great demand	200 requests daily
Device	computer
Wealthy	millionaire

Being specific usually communicates your message clearly. Consider the following:

Draft:

The manuscript was rather large.

Recast:

The manuscript had 1,500 pages.

Draft:

The group leader said she had guided lots of students on camping trips in recent years.

Recast:

The group leader said she had guided 5,000 students on camping trips in the last 10 years.

Avoid Unnecessary Jargon. When you edit your drafts, reconsider what your audience knows about your subject. Don't assume your readers will know the abbreviations, special terms, and definitions you've used. An experienced computer user, for instance, may know terms such as *BIOS*, *video drivers*, and *CPU*, but such terms may be foreign to the novice computer user. Similarly, a geneticist is likely to know terms such as *meiotic anaphase*, *derivative chromosome*, or *diplotene*, but these are likely to be unfamiliar terms to students in an introductory biology class.

When in doubt, define and explain your terms. A short definition following the term helps make the point:

Youth counselors should be concerned about personal liability—in other words, being legally and financially responsible—for youngsters under their supervision.

Each national park has a limited carrying capacity—the maximum number—of visitors a park can handle annually.

Similarly, an abbreviation after the first use of an unfamiliar term helps readers recall what the abbreviation means:

The laser printer prints 13 pages per minute (PPM).

When you revise a document, ask whether your readers will understand the terms and abbreviations you use. When in doubt, define or explain special terms and abbreviations:

Draft:

Some publishers ask their authors to review the bluelines.

Recast:

Some publishers ask their authors to review the page proofs.

A Final Word on Cutting Clutter. Whenever you edit clutter, keep in mind that cutting length usually makes your sentences shorter, presents your ideas more succinctly, and helps communicate your ideas more effectively. Cutting clutter often improves readability, making your documents more understandable for your intended readers. Today's computers and word processing programs make it easy for you to identify and cut clutter.

LOOKING AHEAD

Some of the most powerful tools in today's word processing programs are designed to support editing. Spell Checking tools help you quickly identify misspellings; the **FIND AND REPLACE** command helps you make global changes across even the longest documents; and Grammar and Style Checking tools—if used with appropriate caution and skepticism—help you identify potential problems with grammar and style. These tools can save you a great deal of time on future writing projects. Perhaps most important is knowing that you can use these tools as you near the end of a project, allowing you to focus on more important issues, such as developing strong arguments and supporting them with convincing evidence, right from the start.

chapter 3

Improving Format, Layout, and Document Design

The first thing readers notice about a document is its appearance. In this chapter, we discuss how you can enhance the visual appeal of your documents through the use of a computer. We'll explore why writers choose to invest a great deal of time designing and formatting some documents, while they're content to do little more than dash off others. Then we'll identify key concepts associated with document design, format, and layout, and explore how you can use your word processing program's formatting and layout capabilities to produce attractive, well designed documents. We'll conclude the chapter by discussing the role that typographical style guides can play in harmoniously linking document design elements.

THE FORMATTING DILEMMA

When should you consider formatting, layout, and design of your document? How elaborate should the appearance of your document be? The answers to these questions depend on your audience and purpose, as well as the type of document you're creating. In some cases, you'll want to invest hours giving a document the polished appearance of a professionally-prepared publication. In other cases, you'll be content to produce a document that's nearly indistinguishable from a typewritten letter or report.

Audience and purpose are the first things to consider when asking yourself whether to invest time designing and formatting a document. Are you writing an essay or report for a class? An article for a campus newspaper? A newsletter for a campus or community organization? An internal report for middle level managers? A corporate report for stockholders? In each case, your audience and purpose will vary, from creating a document for a teacher—in which case, the primary needs of your audience will be ease of reading and the ability to make marginal comments—to creating a document for people who have invested their money in your company—in which case, the primary needs of your audience will be to understand the company's financial status and prospects.

Creating a document that a teacher will find easy to read and annotate is not difficult. Typically, you need to select a readable font (for instance, a 12-point Roman face), double space the text, and set up inch to inch-and-a-quarter margins. Teachers sometimes provide guidelines for formatting papers and assign a standard style manual for students to follow (Panel 3.1).

If you're writing a document for publication in a newspaper or magazine, you'll probably be working with an editor who will have needs similar to those of a teacher: a document that's easy to read and edit. Moreover, most newspapers and magazines employ document designers and graphics artists who work with text provided to them by writers and editors. Writers, in this setting, are expected to do little more than provide a basic manuscript that can later be formatted and prepared for publication. Author's guidelines for newspapers and magazines, as a result, usually call for a double-spaced manuscript with illustrations, if any, inserted at the end of the document, as well as the accepted dictionary and major style guide to follow (Panel 3.2). If you're writing for a newspaper or magazine, obtain a copy of the publication's guidelines for authors (see Appendix D for a list

Panel 3.1
Instructor's Assignment Guidelines for Technical Writing

All assignments will be typed. Use a laser printer or ink-jet printer that produces black copy. If you use a dot matrix printer, make sure it has a good, black ribbon. When preparing the assignment follow these guidelines:

- Submit double-spaced, typed copy set on 12-point text
- Use 20-pound bond paper
- Use 1.5 inch margins on all sides
- Use copyediting symbols to correct minor errors
- Correct and reprint dirty copy
- Print a copy of your assignment and retain it in your files
- Type your name, section number, course title, and assignment subject in the upper left-hand corner of the first page
- Drop down half way on the first page before starting to type (this leaves room for written suggestions and critiques)
- Type your name, assignment, and page number in the upper left-hand corner of the second and subsequent pages

Panel 3.2
Sample Publisher's Author's Guidelines

If you're preparing a document that will later be published as an annual report, progress report, technical report, book, brochure, leaflet, journal article, magazine article, or Web page, your document will undergo a content review, sometimes called a substantive edit, and then a style edit to ensure your document conforms to the organizations' style guidelines. Content specialists usually review the content, while an editor conducts the style edit—usually referred to as copyediting. Next, the editor adds type specifications—front, typeface, and size specifications—before your draft document is typeset and prepared for printing. The production process, detailed and costly, entails dozens of steps that move your document from a draft to a polished, printed publication.

Below, we've provided excerpts from the author's guidelines that we worked with while preparing our manuscript for publication.

Sample Publisher's FINAL CHECKLIST for Author's Submitted Manuscripts

Please review the following before mailing your completed manuscript. These guidelines summarize the materials that should accompany your manuscript and the manner in which they should be submitted. For further information on any of the items below, please consult the Author Guide or call your Editorial Assistant.

- Your entire manuscript must also be submitted on **3½-inch diskette.** (In the case of revisions when you are using tearsheeted manuscript, only new material should be submitted on disk.) Please also send a list of the disk files and their contents. **The disk copy must match the printed copy word for word.** Keep a disk copy for yourself.
- Submit **one original manuscript only,** and keep a copy for yourself.
- Text should be **typed, double-spaced, with a minimum 1¼-inch margin** on all sides. *This rule also applies to long quotations and references.*
- **Each original piece of artwork and each table should be submitted on a separate sheet,** not run in with the text. The original artwork should be submitted in a separate folder entitled "Art Manuscript." Tables must be grouped by chapter and placed together at the end of each chapter's text.
- **A photocopy of each piece of art and each table should be placed at the end of the chapter** in which it will appear. Materials should be ordered as follows: Chapt 1 text, Chapt 1 xeroxed tables, Chapt 1 xeroxed art, Chapt 2 text, Chapt 2 xeroxed tables, Chapt 2 xeroxed art, Chapt 3 text . . .
- Make sure each piece of art and each table has a **label** (double-numbered, i.e., Figure 1-2 is Chapter 1, Figure 2) and that each is called out by this label in the text.

continued

Panel 3.2 *(continued)*

- **Figure captions** should be submitted as a separate list, grouped at the end of each chapter's text.
- **Frontmatter** should appear in the following order: Title Page, Dedication, Contents, Foreword, Preface, Acknowledgments.
- **Page numbering** should begin on Page 1 of Chapter 1 with the number 1 (do not number frontmatter). Then, each page is numbered consecutively, *including all pages that bear photocopies of art and tables.* This numbering system is used to ensure that no pages of your manuscript are lost, missing, or out of order.
- **Any material designated for the margins** of the finished text (i.e., marginal glossary, marginal critical thinking questions, etc.) should be typed on a separate sheet and placed at the end of the text of the corresponding chapter. All items must also be keyed to call-out points in the manuscript that indicate where in the margin each item should appear.
- Finished manuscripts should bear **no handwritten changes.** if you must add handwritten correction, please do so in black *pen* and indicate if these changes were made on the disk copy.
- **Tearsheet manuscript** will not be accepted unless the left and right sides of the tearsheet have been completely taped down with clear tape onto 8½ × 11 inch paper. Double-column tearsheet must be split down the middle, so only one column is taped on each sheet of paper.
- Your **final manuscript package** should be made up of the following components:

Text	Author's Suggestion Form
Art Manuscript	Disk Copy of Text and Art
Tables	Author's Checklist
Permissions	List of Photograph Specifications

obtain a copy of the publication's guidelines for authors (see Appendix D for a list of commonly accepted style guides and dictionaries).

When you write papers for a class, your design considerations seldom extend beyond the use of a readable typeface (e.g., Roman or Courier) and adequate margins. Similarly, when you write for a magazine or newspaper, your editor will expect nothing more than a double-spaced set of pages with enough room to make edits. In other situations, however, your audience and purpose will require you to spend much more time considering document design:

- If you're working for a campus or community group, you might write and design a wide range of documents, among them newsletters, brochures, letters, reports, and Web pages. In many smaller organizations, you'll be asked to take on the multiple roles of writer, editor, graphics artist, and designer.

- If you're in sales, you might need to prepare a wide range of reports and proposals for management. Your ability to communicate clearly with management will have bearing on your success in obtaining their support for your initiatives.
- If you're working in corporate communication at a small company or agency, you might be required to be a one-person production team, handling many or even all elements of document design and preparation.

If you find yourself spending more and more time designing and producing documents, you'll most likely want to learn how to use desktop publishing software such as Adobe Pagemaker, Quark QuarkXPress, Microsoft Publisher, Corel Ventura, or Adobe Framemaker. Desktop publishing software offers you a much wider range of options for document design and preparation than can be found in most word processing packages.

We don't want to give the impression, however, that word processing software is of no use in document design. For the vast majority of documents, word processing programs offer more design, layout, and formatting capabilities than you'll ever need. In the next section, we discuss some of the key concepts in document design, layout, and formatting. We conclude the chapter with a detailed discussion of how your word processing program can help you create effective and attractive documents for a wide range of audiences and purposes.

ELEMENTS OF DESIGN

To design documents, begin with an analysis of your audience, purpose, and resources. We've addressed the issues of audience and purpose in the previous section, so we'll turn here to a discussion of resources. Resources include things like your knowledge of document design, your budget (if any) for printing and related services, your access to graphics services, and technological support such as your computer, software, and printer. Each of these resources plays a role in shaping the design of your document.

Your knowledge of document design is the most important resource issue you face. Our goal in this chapter, of course, is to help broaden your knowledge of document design, with specific attention to how computers can support design, layout, and formatting. Knowledge of issues such as the relative legibility of certain font faces and styles, for instance, can help you design more effective documents. Similarly, knowledge of the wide range of document types that can be easily prepared on a computer can help you tailor a particular document more effectively for your audience.

Your budget for printing and production, as well as your access to graphics services, will play an important role in the design of your document. If you're creating a document that will go to one or two people, then production costs are less of an issue. If you plan to produce hundreds or thousands of copies of a document, however, costs of production will have a strong impact on your design decisions. Do you have funds for full-color printing? For one or two colors? For black and white? Does funding limit the number of pages you can create? If so, you may have to cut content or graphics or perhaps reduce the size of your body font. Can you count on outside services, or a friend, to help with graphics issues, such as photos and drawings? Or will you have to rely on your own skills, or clip art, to add graphical interest to your publication?

Finally, do you have the technological resources to create the document yourself? Or will you have to send the document elsewhere? For the production of small numbers of high quality documents you don't need to own a high-powered computer and color printer—you can simply visit a retail establishment such as Kinko's. If you're looking at a large production run, on the other hand, you'll probably want to work with a commercial printer. Each of these technological resource issues, from your computer hardware and software to the capabilities of your printer and photocopier, affects your decisions about the design and production of your document.

Each of these resource issues will have an impact on the design of your document. Although we can't help you with funding and technology resources, we can help enhance your knowledge of document design. Below, we focus on the major document design elements: page size and orientation; typography; margins, columns, and gutters; line spacing, alignment, paragraphing, and tabs; heads; numbered and bulleted lists; borders and colors; figures and tables; page numbers; headers and footers; and tables of contents, indexes, and glossaries.

Page Size and Orientation

Page size and orientation are the foundation on which print documents are created. In the United States, most documents use standard 8.5 × 11 inch paper with text printed in portrait, or vertical format. Word processing programs typically use this as their default page setup (Figure 3.1). There are good reasons for using this page size and orientation for most documents, among them readers' familiarity with the format and the relative ease of printing and distributing documents that use this page format.

Consider departing from the norm, however, if you are creating a document that would be better served by another format. Use an 11 × 17-inch format, for instance, if you want to create a newsletter that is folded, rather than stapled—you'll end up with two pages on each side of a sheet of paper. Use an 8.5 × 14-inch format if you are creating a tri-fold brochure—you'll end up with three pages on each side of a sheet of paper. Or, if you're working with a commercial printer, explore the range of paper sizes they offer and adjust your document's page size and orientation accordingly.

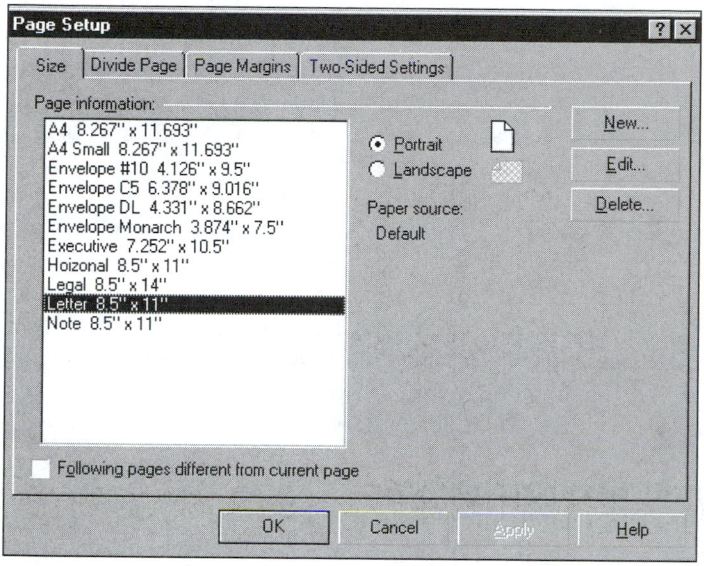

FIGURE 3.1 Page setup dialogue box

Keep in mind the distinction between page size and paper size when you are designing a document. You'll need to specify each in your word processing program's Page Setup or Page Properties dialogue box (Panel 3.3).

Panel 3.3
Page Setup

Most word processing programs allow you to format a document by using the PAGE SETUP or DOCUMENT PROPERTIES command on the File menu. Use the Page Setup dialogue box to specify the size of the paper you want to use for your document, page orientation (landscape or portrait), and page margins. Many word processing programs also allow you to specify additional settings such as columns, headers and footers, and two-sided printing. You can apply settings to an entire document, to sections of a document, or to individual pages.

HOW TO FORMAT PAGES

1. Click on the File menu
2. Click on PAGE SETUP or DOCUMENT PROPERTIES
3. Adjust settings in the dialogue box

Typography

The development of high quality printers and operating systems using graphical user interfaces (GUIs), such as the Apple Macintosh OS and the various versions of Microsoft Windows, has increased our ability to consider typography in document design. Most word processing programs support a wide range of fonts, sizes, and other characteristics.

Font refers to a family of type with all of its letters, numbers, punctuation and other typographical symbols (Figure 3.2). Common fonts include Times Roman, Courier, and Arial. Font style refers to regular body text, italic text, bold text, and bold italic text. Fonts are generally classified as either having serif or sans serif typefaces. Serif faces have small brush strokes or fine lines at the end of the stroke, as in Times Roman type, while sans serif typefaces lack the serifs—the brush strokes or fine lines (Figure 3.3). Effects refers to superscript, subscript, strikethrough, small caps, and all caps.

Size refers to the point size of type. Type is measured in points with 72 points to an inch. The ascenders and descenders give different fonts a complicated estimating point size (Figure 3.3). When type size is considered, it is either body

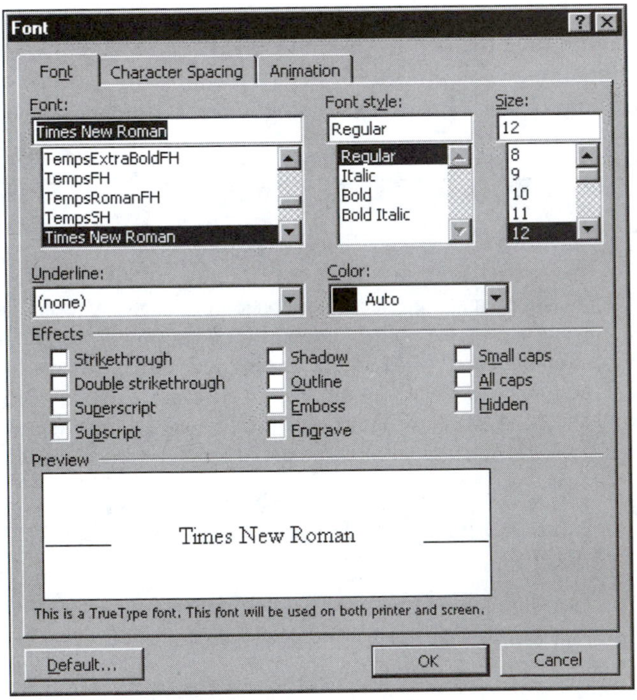

FIGURE 3.2 Font selection dialogue box

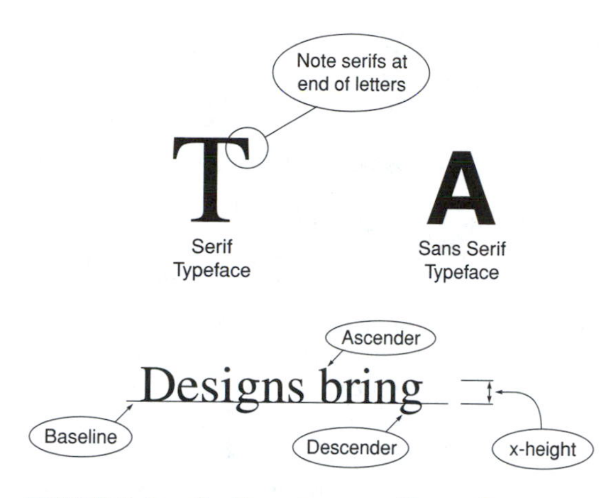

FIGURE 3.3 Serif and san serif typefaces, x-height, ascenders, and descenders

text, usually 12-point type or smaller, or headline text, usually 14-point type or larger.

Printers and publishers have developed typographical guidelines based on years of experience and research. Below, we discuss key aspects of those guidelines.

Point Size and Leading. Research suggests "safe zones" for different type size, leading, and line length. Many of today's magazines, books, journals, and other publications now follow the guidelines that emerged from Miles Tinker's (1963) summary of early research on this issue. Publishers usually set the body copy—the narrative of a document—in 9-point to 12-point type. Select 12-point type and you'll produce a document with about six lines of type per column inch of the column. Select 9-point type, and you'll produce a document with about eight lines of type per column inch.

In contrast, publishers usually set headline copy—or heads—in 14-point to 72-point type. Heads function as markers or street signs that guide readers through a document. Thus, heads should be easy to spot, read, and understand. However, using type that is too large, such as 48- to 72-point heads, seldom enhances your documents. Similarly, avoid using type that is too small in comparison to your body type.

Although you can set body copy at any point size, consider your readers and where they are reading your document. For example, the American Association of Retired Persons (AARP) uses larger point sizes because AARP recognizes that many chronologically advantaged people have trouble reading small type, as their eyes cannot focus as easily as they once did. Or consider the situation where documents will be read under low light conditions, such as in a car at night.

Selecting an appropriate point size becomes more complicated with different fonts. On casual observation, 11-point type in one font may appear larger than in another font. For example,

11-point Universe versus 12-point Times Roman

The ascenders and descenders, the brush strokes above and below the line, can change the appearance of type size. Further complicating the appearance of text is the spacing, or leading, between lines of text. When type was set with metal versus today's desktop publishing systems, pieces of lead were inserted between lines of type to give the desired appearance; thus the text could be spaced out to give different appearances (see Figure 3.4).

In Figure 3.4, the left example is 12-point Times Roman with normal spacing (i.e., 12-point base), while the right example is 12-point Times Roman with six more points of spacing or leading. The additional spacing between lines opens the text, but may hamper reading.

Body and Head Fonts. Generally, body copy—the copy that reports your information—is set in a serif typeface, such as Times Roman. Research has not provided a definitive answer as to whether serif or san serif typefaces are more legible (Schriver 1995), but the majority of publications in the United States use serif typefaces for body copy.

In contrast to body copy, many publications use both serif and sans serif typefaces for heads. Typically, readers quickly scan heads and, if interested, read

Mountain lions roam the foothills west of Fort Collins, and sometimes wander into the city. In fact, the Colorado State University campus patrol cornered a mountain lion on campus about 2 a.m. one August morning in 1987.	Mountain lions roam the foothills west of Fort Collins, and sometimes wander into the city. In fact, the Colorado State University campus patrol cornered a mountain lion on campus about 2 a.m. one August morning in 1987.
12-point type on 12-point base	**12-point type on 18-point base**

FIGURE 3.4 A comparison of leading

the associated body text. Heads serve a function similar to street signs—they tell readers what's coming and give a general overview of information in a document.

Limiting Fonts, Font Styles, and Underlined Text. In many word processing programs you have the opportunity to select from several dozen fonts, and you can purchase hundreds of additional fonts should you want additional variety for typographical design. But having a wide range of fonts does not mean that you need to use them in the same document. Graphic designers have long argued that you need to provide consistency in design, something that is undermined if you choose to use a wide range of fonts. Following conventions in the publishing industry, we encourage you to limit the number of fonts you use in a given document to no more than two to four.

Similarly, use font styles and underlining with restraint. Research generally suggests readers find and prefer regular body copy to lengthy amounts of boldface, italics, and capitalized text (Tinker 1963, Felkner et al. 1982). Extensive underlining—seldom used in printed documents prior to the advent of word processing programs and laser printers—also presents comparable problems for readers (Figure 3.5). Although no hard rule has emerged, prudent document designers avoid more than a line or two of text in all boldface, italics, capitals or

Mountain lions roam the foothills west of Fort Collins, and sometimes wander into the city. In fact, the Colorado State University campus patrol cornered a mountain lion on campus about 2 a.m. one August morning in 1987.	*Mountain lions roam the foothills west of Fort Collins, and sometimes wander into the city. In fact, the Colorado State University campus patrol cornered a mountain lion on campus about 2 a.m. one August morning in 1987.*
MOUNTAIN LIONS ROAM THE FOOTHILLS WEST OF FORT COLLINS, AND SOMETIMES WANDER INTO THE CITY. IN FACT, THE COLORADO STATE UNIVERSITY CAMPUS PATROL CORNERED A MOUNTAIN LION ON CAMPUS ABOUT 2 A.M. ONE AUGUST MORNING IN 1987.	<u>Mountain lions roam the foothills west of Fort Collins, and sometimes wander into the city. In fact, the Colorado State University campus patrol cornered a mountain lion on campus about 2 a.m. one August morning in 1987.</u>

FIGURE 3.5 Examples of boldface, italics, all capitalized, and underline text

Panel 3.4
Formatting Characters

Formatting characters is typically accomplished by using the FORMAT FONT command. Depending on your word processing program, you can specify:

- Font face, such as Courier or Times Roman
- Font style, such as normal, bold, italic, all capitals, small capitals, or bold and italic
- Font size (typically referred to as point size), such as 10-, 12-, or 48-point
- Font color
- Font effects, such as superscript, subscript, underlining, shadow, and strike-through

You can format fonts using menu commands, keyboard commands, or by clicking on a button bar.

HOW TO FORMAT FONTS

Using the Menu

1. Select the text you want to format
2. Click on the Format or Text menu
3. Click on FONT
4. Adjust settings in the Font or Text Properties dialogue box

Using the Button Bar

1. Select the text you want to format
2. Click on the TEXT PROPERTIES or FORMAT FONT icon in the button bar

Using the Right Mouse Menu

1. Select the text you want to format
2. Right click with your mouse
3. Click on FONT or TEXT PROPERTIES
4. Adjust settings in the dialogue box

underlined. Use boldface, italics, and capitalized text for selected emphasis, rather than over-lengthy passages—it's much like crying "wolf" too many times.

Margins, Columns, and Gutters

Margins are the white space that borders text on a page. Columns are lines of text. In the vast majority of documents produced using word processing programs, most pages contain a single column. Multiple columns, in contrast, are typically used in publications such as newspapers, magazines, and newsletters. The white space between columns in documents with multiple columns is called the gutter.

Word processing programs allow you to control the width of margins, columns, and gutters in several ways (Panel 3.5). First, you can set up your pages using the page setup or document properties dialogue box. Second, you can use the ruler that can be displayed at the top of your document. And third, you can use specific commands on the menu bar (such as **FORMAT COLUMN**) or button bars.

If you're producing a document that will be reviewed, edited, and critiqued by others, use one- to one-and-a-half inch margins, a single column, and double space the copy. This will give reviewers, editors, and instructors plenty of space for adding comments and edits. Although many reviewers, editors, and instructors are beginning to comment online, many others prefer to write by hand in the margins.

If you are producing a document in its final form, use margins, columns, and gutters to make the document more effective and visually attractive. Too little white space and the document will appear dense, potentially discouraging readers. Too much white space and the text will appear lost on the page.

Panel 3.5
Columns

Columns are found in many documents. Leading word processing programs let you create columns in your documents by clicking on an icon in a button bar or by using menu commands. You can usually locate column formatting commands on the Format or Text menus.

In some word processing programs, you can also quickly create columns by using a button bar icon. Simply select the text you want to format in columns, click on the icon, drag until you've selected the desired number of columns, and release the mouse button (Figure A).

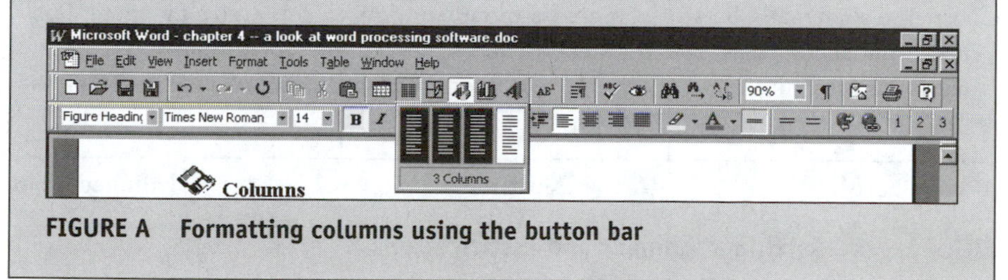

FIGURE A Formatting columns using the button bar

Line Spacing, Alignment, Paragraphing, and Tabs

Line spacing is a term inherited from the time in which typewriters were the most commonly used tool for writing. Single spacing referred to the practice of not placing any extra spacing between lines on a typewritten page, while double and triple spacing referred to the practice of inserting one or two extra lines respectively between typed lines of text. Word processing programs have preserved the convention of referring to extra spacing between lines as single and double spacing. In addition, most word processing programs include a one-and-a-half line spacing option.

Alignment refers to the appearance of text relative to the left and right sides of a column of text (Figure 3.6). In most word processing programs, text can be aligned right, left, centered, or justified. Left alignment produces a column of type in which the words align on the left-hand side of the column, but create a ragged right margin—in other words, the lines end in full words but the words do not align on the right-hand side of the column. Right alignment produces a column of type in which the words align on the right side, but create a ragged left margin. Centered alignment produces a column of type in which the words align along a center line with ragged right-hand and left-hand margins. Justification fills each individual line so that the beginning and last letters in each line align vertically, usually by adding extra space between words and/or letters in the line of text.

Paragraphing refers to the way you control the format of paragraphs (Panel 3.6). Typical formatting commands at the paragraph level include line spacing, line indentation, text alignment, spacing before and after paragraphs, and special formatting commands such as hanging indents.

Mountain lions roam the foothills west of Fort Collins, and sometimes wander into the city. In fact, the Colorado State University campus patrol cornered a mountain lion on campus about 2 A.M. one August morning in 1987.	Mountain lions roam the foothills west of Fort Collins, and sometimes wander into the city. In fact, the Colorado State University campus patrol cornered a mountain lion on campus about 2 A.M. one August morning in 1987.	Mountain lions roam the foothills west of Fort Collins, and sometimes wander into the city. In fact, the Colorado State University campus patrol cornered a mountain lion on campus about 2 A.M. one August morning in 1987.
Aligned Left	**Justified**	**Aligned Right**

FIGURE 3.6 Setting alignment and justifications

Panel 3.6

Formatting Paragraphs

Like characters, paragraphs can be formatted in several ways. Depending on your word processing program, you can specify:

- Alignment (left, right, centered, or justified)
- Line spacing (single-spaced, double-spaced, etc.)
- Indentation from the right and left margins
- Special formatting such as hanging indentation (used to indent text below the first line)
- Space before and after the paragraph

To format paragraphs, use the Format Paragraph or Text Properties menu commands. You can also access paragraph formatting commands by right-clicking with your mouse on a paragraph.

The most common paragraph indentation style is a five-space indentation at the beginning of each paragraph. Although the five-space indentation is common in double-spaced documents, some guidelines for single-spaced documents call for double spaces between paragraphs and no paragraph indentations.

Tabs are pre-set locations used to align text across a line of type. Most tabs are set every five characters or letters across a line. By establishing preset tabs, you can align quotations and paragraph indents. Most word processing programs allow you to set several kinds of tabs:

- Left tabs, which serve as a left margin for the tabbed text
- Right tabs, which serve as a right margin for the tabbed text
- Center tabs, which serve as a center point for tabbed text
- Decimal tabs, which align text (usually numeric text) along the decimal point
- Leader tabs, which create a leader (usually a series of dots, a line, or a broken line) next to the tabbed text

tip 3.1 Avoid Excessively Deep Paragraphs

White space plays a vital role in making documents both easier to read and visually attractive. In today's information-saturated world, newspapers, magazines, journals, book publishers, and Web pages use typographical styles with more

continued

white space. One technique that helps infuse more white space into a document is using shorter, succinct paragraphs. They help break up text.

Not only do shorter paragraphs break up the text, they make it easier for readers to find key information. In many functional documents—documents providing information and instructions such as computer manuals, instruction sheets, and specifications—the key is making information easier for readers to find. Consider too newspapers. Over the last 75 years newspapers have evolved from those using dense typographical designs to ones employing more open and visually appealing publications, such as *USA Today*. In today's society, providing visually pleasing and easy-to-locate information is a critical design factor for functional documents.

When designing your document, consider the conventions followed in similar publications. Note especially the paragraph length and white space and then follow a similar design for your document. For example, paragraph depth in newsletters ranges from 6 to 10 lines of text. If your document is a newsletter, then follow that convention. In contrast, other documents, such as novels, may use lengthy paragraphs depending on the author's personal style. Novels often have paragraphs running 10 or more lines deep. If your document is a novel, then follow that convention.

Heads

Heads and subheads signal major divisions of a document to your reader. Sometimes heads are called headings or headlines. Heads and subheads provide a visual break and let readers know something about the content of the following section. Heads are typically used in levels: number one heads, number two heads, number three heads, number four heads, and so forth.

Each head has a typographical style. For example, the sample publisher's author's guidelines call for number one heads to be centered and capitalized; number two heads to be centered using upper case, lower case; number three heads to be flush left with upper case, lower case; and number four heads to be indented, upper case, lower case, in italics, and followed by a period (Figure 3.7).

Word processing programs allow you to specify heads by using the style command. Styles are sets of type specifications such as font face, font size, font style, font effect, alignment, and indentation (Panel 3.10). If you format text using styles, you can easily change the appearance of all text using the style by changing the type specifications for the style. This allows you to easily change and review the formatting in your document—a real advantage as you work with the design of your document.

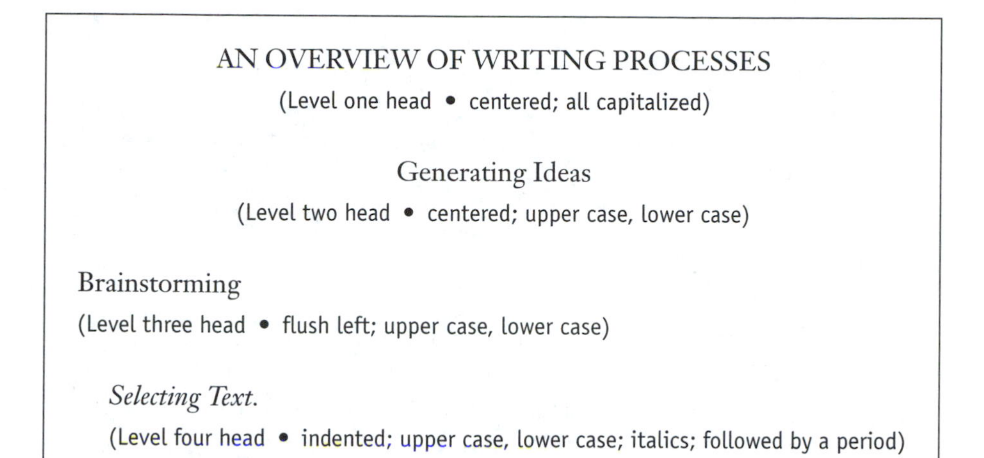

AN OVERVIEW OF WRITING PROCESSES

(Level one head • centered; all capitalized)

Generating Ideas

(Level two head • centered; upper case, lower case)

Brainstorming

(Level three head • flush left; upper case, lower case)

Selecting Text.

(Level four head • indented; upper case, lower case; italics; followed by a period)

FIGURE 3.7 Sample publishing levels of heads

Numbered and Bulleted Lists

Numbered and bulleted lists can enhance the appearance of your documents and convey information concisely and effectively. Numbered and bulleted lists help break up text. They can also provide readers with a quick overview of topics to come. Word processing programs allow you to use Arabic or Roman numerals in numbered lists and a variety of typographical symbols in bulleted lists (Panel 3.7). Use numbered lists when you need to make points in sequence, such as instructions, and use typographical bullets for more general listings.

Borders and Colors

In addition to the textual elements we've discussed so far, you can use nontextual elements to add interest and attractiveness to your documents. Most often, you'll use these elements to call attention to particular parts of your documents. Sometimes, you'll use them because they simply look good. Whatever your reason for using borders and colors, consider the impact they will have on your reader.

Leading word processing programs allow you to easily add borders to your documents. You can create boxes around pages or sections of text with varying line weights, shading, and spacing. You can run lines along one or more of your margins. And you can add three-dimensional effects. If you are printing in color or creating documents that can be read online, you can change the color of text, background, lines and boxes, and shading. To learn how your word processing program allows you to use color and borders, consult online help.

Panel 3.7
Bulleted and Numbered Lists

Bulleted and numbered lists are found in a variety of documents and word processing programs have added simple commands for creating them. Bulleted and numbered lists help writers present information concisely and attractively. Depending on the information, readers may find it easier to follow information presented in a bulleted or numbered list than in paragraph format.

To create a bulleted or numbered list, select the text you would like to transform into a list and then click on a bullet list or number list icon on the button bar. You can also access bullet and number commands through the Insert, Text, or Format menu in most word processors.

You can combine bulleted and numbered lists and present them in different levels, as you do in an outline. Typically, you can modify the appearance of your bulleted and numbered lists through a dialogue box, either by right-clicking with your mouse on selected text and clicking on the appropriate command or by accessing menu commands (Figure A).

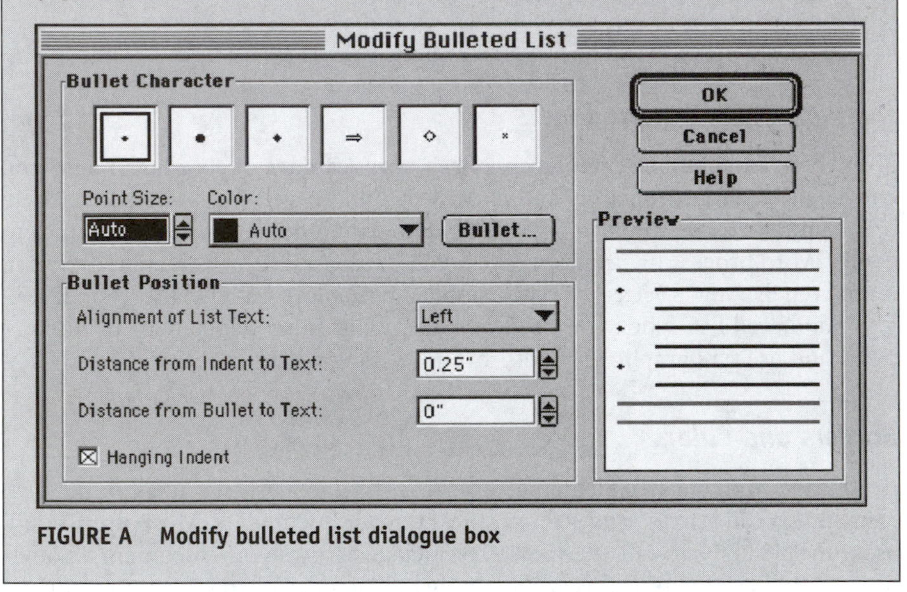

FIGURE A **Modify bulleted list dialogue box**

The primary concerns related to adding color to a document are whether you can print the document in color and whether it will add to production costs. If these are not an issue for you—for instance, if you have access to a color printer or you are creating the document for online distribution—then by all means explore the role color can play in your documents. You will find that it adds a great deal to their visual appeal; perhaps most important, color provides a powerful means of directing readers' attention to particular parts of a document.

Panel 3.8
Footnotes and Endnotes

Footnotes and endnotes are another useful document design device. Footnotes and endnotes are useful for indicating sources of information for a document, related information that you would prefer not to present in the body of a document, and acknowledgments of contributions to a project or document. In draft stages, you can also use footnotes and endnotes as annotations to remind yourself about things you need to do in subsequent drafts, for example.

Most word processing programs allow you to easily add footnotes and endnotes to a document. As with page numbers, most word processing programs provide a range of options for formatting and controlling the placement of footnotes and endnotes. In most word processing programs, you can access the **FOOTNOTES AND ENDNOTES** command via the Insert or Create menu.

Like many other document design elements, such as font faces and styles, you can overuse borders and colors. Be cautious, as a result, in your use of borders and colors. Consider where they will be most effective in helping you meet your goals for your document, then use them sparingly. You'll find, as in most things, that judicious use of colors and borders goes a long way.

Tables and Figures

In business, sciences, and engineering, among other fields, tables and figures play important roles in documents. Tables usually present numeric data, such as percentages, scores, and statistical results, although they occasionally present textual information. Figures, in contrast, present charts, graphs, drawings, photographs, equations, and more (Figure 3.8). When you think of tables and figures, consider the adage, "*A picture is worth a thousand words.*" Because tables and figures allow you to present a great deal of information in a small space, they are extremely useful document design elements. Remember, though, that they draw attention to themselves and away from text. When readers see a table or figure on a page, they often look there first. If you would rather have them focusing on something else in your document, reconsider their use.

Word processing programs allow you to easily insert, and even to create, charts and graphs. They also provide the ability to insert and format photographs and other graphical materials, such as clipart and graphics files. For information about how your word processing program supports these functions, consult online help. For more information on creating tables, see the discussion in Chapter 1, Panel 1.3.

TABLE 1.1 Average IBM and Macintosh experience in months of Colorado State University Students, fall 1996 (n = 343 students)

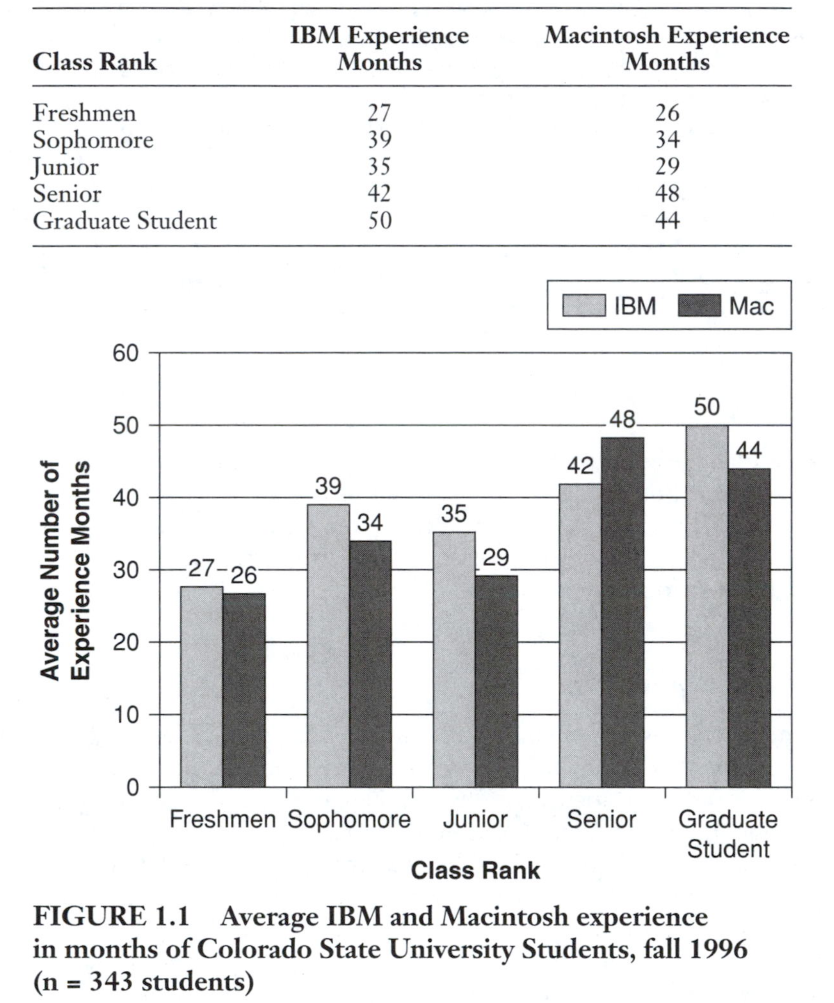

Class Rank	IBM Experience Months	Macintosh Experience Months
Freshmen	27	26
Sophomore	39	34
Junior	35	29
Senior	42	48
Graduate Student	50	44

FIGURE 1.1 Average IBM and Macintosh experience in months of Colorado State University Students, fall 1996 (n = 343 students)

FIGURE 3.8 Illustration of data presented in both table and figure format

Page Numbers

Adding page numbers to a document is extremely easy. You can specify location (e.g., top or bottom of page, right, left, or centered), style (e.g., Roman or Arabic numerals), and size, among other options. The program will then insert page

Panel 3.9
Online Information about Desktop Publishing

The Internet—and the World Wide Web in particular—offers a range of information sources on desktop publishing. Good places to start include the directories on Yahoo <http://www.yahoo.com>, a leading Web search site (see Chapter 4). Search Yahoo, or other search engines, such as Lycos <http://www.lycos.com> and Excite <http://www.excite.com>, using the phrase *desktop publishing.* Some of the category listings on Yahoo include:

- BUSINESS AND ECONOMY:COMPANIES:COMPUTERS:SOFTWARE:DESKTOP PUBLISHING
- COMPUTERS AND INTERNET:DESKTOP PUBLISHING
- COMPUTERS AND INTERNET:SOFTWARE:REVIEWS:TITLES:DESKTOP PUBLISHING
- COMPUTERS AND INTERNET:SOFTWARE:DESKTOP PUBLISHING

numbers automatically. You can also choose to restart or continue numbering in successive sections. This allows you to use Roman numerals in the front matter of a document, and Arabic numerals in the rest.

To access the page number commands in most word processing programs, consult online help. Location of the page number menu commands varies widely across major word processing packages.

Headers and Footers

In addition to page numbers, most word processing programs enable you to add headers and footers to documents (Figure 3.9). As the names imply, headers run along the top of pages in a document, while footers run along the bottom. Headers and footers can contain text and automatic page numbers, among other options. In addition to labeling documents, headers and footers can identify whether a document is a draft or a final version and can indicate its file name and location.

If you're working on a document, consider adding a footer that includes the date, file name, and location on your hard drive. This will save time should you need to locate the document later.

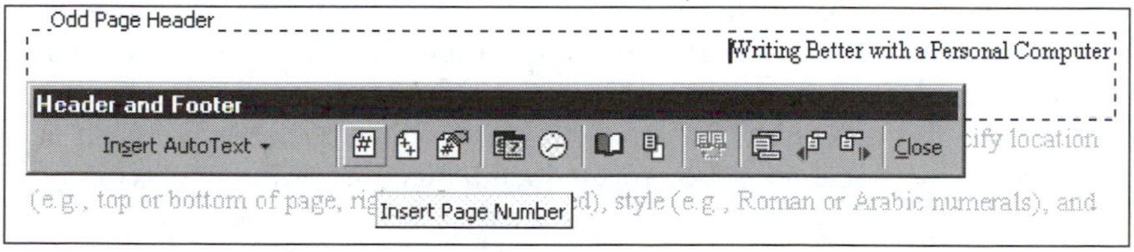

FIGURE 3.9 Header and footer dialogue box

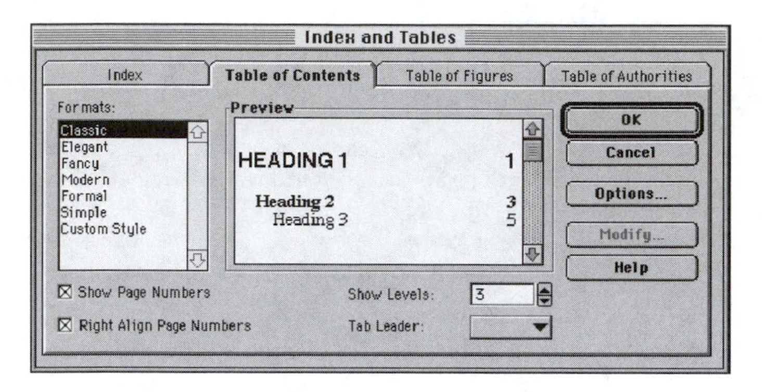

FIGURE 3.10 Index and tables dialogue box

Tables of Contents, Indexes, and Glossaries

Longer documents, such as books, reports, and some proposals, often contain tables of contents, indexes, and glossaries. They are an important part of longer documents, enabling readers to more easily locate information within a document. As such, you should consider them in your design of longer documents.

If you've ever created these document elements manually, you know how much work it involves. When you routinely work on long documents, you'll come to appreciate the speed and simplicity of word processing tools that help you create tables of contents, indexes, and glossaries (Figure 3.10). Equally important, you'll appreciate the flexibility word processing programs offer for controlling the layout and appearance of these elements. As you consider your design of longer documents, explore the tools offered by your word processing program for creating tables of contents, indexes, and glossaries. You're likely to find, as we have, that investing the necessary effort to make them work (chiefly, using the STYLES command, see Panel 3.10) will pay off in a significant saving of time.

DEVELOPING A TYPOGRAPHICAL STYLE GUIDE

In the previous section, we outlined some of the major elements you can consider as you design a document. Most writers are content to work with documents one at a time in a relatively unsystematic manner. When they need a new style, they create one and, if necessary, they change some of the styles they've been using elsewhere in the document. For most documents, and for most writers, this works well.

Consider developing a typographical style guide, however, if you find yourself working on a project where strong design is absolutely necessary to the suc-

Panel 3.10
Styles and Document Templates

One of the most powerful techniques offered to writers by word processing programs is the ability to define and apply styles in a document. In word processing, a style is a set of formatting commands. A style named "Heading 1," for instance, might tell the word processing program to format text as Arial font, bold font style, 18-point font size, underlined, left paragraph alignment, 18-point spacing before the paragraph, 12-point spacing following the paragraph, and single line spacing.

Applying each of these formatting commands individually would take some time even for someone with advanced word processing skills. By using the STYLE command, however, you can apply all of these formatting commands at once. More important, if you decide to change the formatting of a particular style, all the text formatted with that style will change automatically. Styles provide greater control over the appearance of a document, allowing you to experiment easily with a variety of formatting choices.

Word processors that support the STYLE command provide built-in styles, but you can define your own styles or modify existing ones. You can also copy styles from one file to another or store them in the templates used to create new files.

Document templates typically contain a set of styles and other elements that define the way you work with documents based on that template. A template for a Web document, for instance, might contain a button bar that displays commands related to working on the Web, such as opening a Web browser or creating a hypertext link. If you work with the same kind of documents on a regular basis, you can create a template containing styles for that kind of document. To learn more about how to use, modify, and create templates in your word processing program, consult online help.

Whether or not you choose to use document templates, consider using styles when you work with long documents, or with sets of related documents, because of their advantages over formatting text one character, word, or paragraph at a time. You'll find it useful to experiment with style commands to speed your writing.

HOW TO USE STYLES

Using the Menu

1. Select or place your cursor in the text to which you want to apply a style
2. Click on the Format or Text menu
3. Click on STYLES
4. Select the style

Using the Button Bar

1. Select or place your cursor in the text to which you want to apply a style
2. Click on the STYLE icon in the button bar
3. Select the style

cess of your document, or the length of your document is such that you want support for ensuring consistency across the document. A typographical style guide—a set of decisions about how to format and layout text, tables, and figures, as well as how to use colors, borders, and shading—is a superb resource in any major writing project. For that matter, it's a great resource even on smaller projects, if you choose to use it for several shorter texts over a period of time.

The value of a typographical style guide is two fold. First, creating the guide forces you to focus on the details of document design. You'll need to consider font face and styles, styles for heads and subheads, borders and captions for tables and figures, and whether and where to use color. Considering each of these elements in relation to each other will help you create a balanced, well-designed document. Second, the creation of a typographic style guide—and the styles that you create to implement your guide—will allow you to easily modify your document as you create it. The ability to use the **STYLE** command to make immediate changes across a long manuscript is a powerful tool for adapting your document design to your audience, purpose, and resource constraints. If you've planned, for instance, to print a document in color only to learn that you can no longer afford to do so, you'll be able to easily and quickly change many elements of your document by using the style command: the heads and subheads that you defined as using a certain color can be modified for black and white printing with a few mouse clicks in a dialogue box.

In the end, document design has a simple goal: to increase the effectiveness and readability of your document in light of your audience, purpose, and resource constraints. Thinking carefully about typography and the other elements of document design can play an important part in the ultimate success of your writing projects.

LOOKING AHEAD

In the end, document design has a simple goal: to increase the effectiveness and readability of your document in light of your audience, purpose, and resource constraints. Thinking carefully about typography and the other elements of documents design can play an important part in the ultimate success of your writing projects.

chapter 4

Gathering Information Online

A few years ago, teachers who assigned research papers went out of their way to make sure that their students found enough sources. Locating relevant books and articles for a report was a challenging task, and many students had difficulty finding enough useful sources in local libraries.

Today, teachers are more concerned with what students do with the sources they've found. And a key strategy that most teachers address during class is how to sift through the many sources located during searches of the Web, databases, and online library catalogs.

The amount of information available online has grown with astounding rapidity in the past few years, and the pace of that growth is increasing. The local library is now only one of many sources of timely and relevant information. For writers with access to the Internet, the issue isn't what they'll find, but where to look first and what to do with the results of their search.

In this chapter, we'll discuss the wide range of information available online. We'll focus on locating online information sources, searching and browsing online information, evaluating online information, and organizing the information you've decided to use.

LOCATING ONLINE INFORMATION SOURCES

These days, most people think about the World Wide Web when they think about online information. But the Web is only one of many online information sources. Some of the most important, such as databases and online library catalogs, aren't necessarily located on the Internet. Others, such as Gopher, newsgroups and mailing lists, online document collections, chat channels, and MOOs actually predate the development of the World Wide Web. Those observations aside, you can locate almost all of these sources of information via your Web browser and its associated communication tools. In this section, we'll explore each of these sources of information, calling attention to what they are and how you can use them to support your writing.

Online Library Catalogs

Most research libraries have replaced their card catalogs with online catalogs. Online catalogs serve the same purpose as card catalogs, but are easier to search and support multiple ways of locating information. Most online catalogs allow you to search for keywords (specific words or phrases such as "constitution" or "liberty for all"), authors, titles, and subjects. You can often limit your search (restrict your search so that you are not searching the entire holdings of the library) by publication year or subject. You'll read more about search strategies later in this chapter.

Online catalogs supply two types of information about a source—publication and circulation. Publication information can include author, title, edition, publisher, publication city and date, and sometimes a brief description of the contents of the source. Circulation information can include call number, location within the library, whether the source is available, and when it was acquired by the library.

Online catalogs typically serve single libraries or library systems. The SAGE system at Colorado State University, for example, provides information about the holdings of libraries in the Colorado State University system. SAGE is available via the Web (see Figure 4.1) and via Telnet (a character-based program using keyboard commands, see Figure 4.2). SAGE allows you to search for information on library holdings. It also provides access (via the Web or Telnet) to a large number of online databases subscribed to by the University. In addition, you can use SAGE to access interlibrary loan and to contact subject area librarians.

Panel 4.1
What Is the Internet?

The Internet is a network of computers connected to each other by telephone lines, high-speed data lines, microwave relays, and satellite links. Sometimes referred to as the "network of networks," the Internet was developed by the United States to support communication among computers in the event of a national catastrophe, such as a nuclear attack.

The Internet existed for many years before the World Wide Web was developed. Until the World Wide Web, users of the Internet accessed files on other computers using terminal programs such as Telnet and FTP (File Transfer Protocol). More recently, Gopher was developed to simplify access to files on computers connected to the Internet.

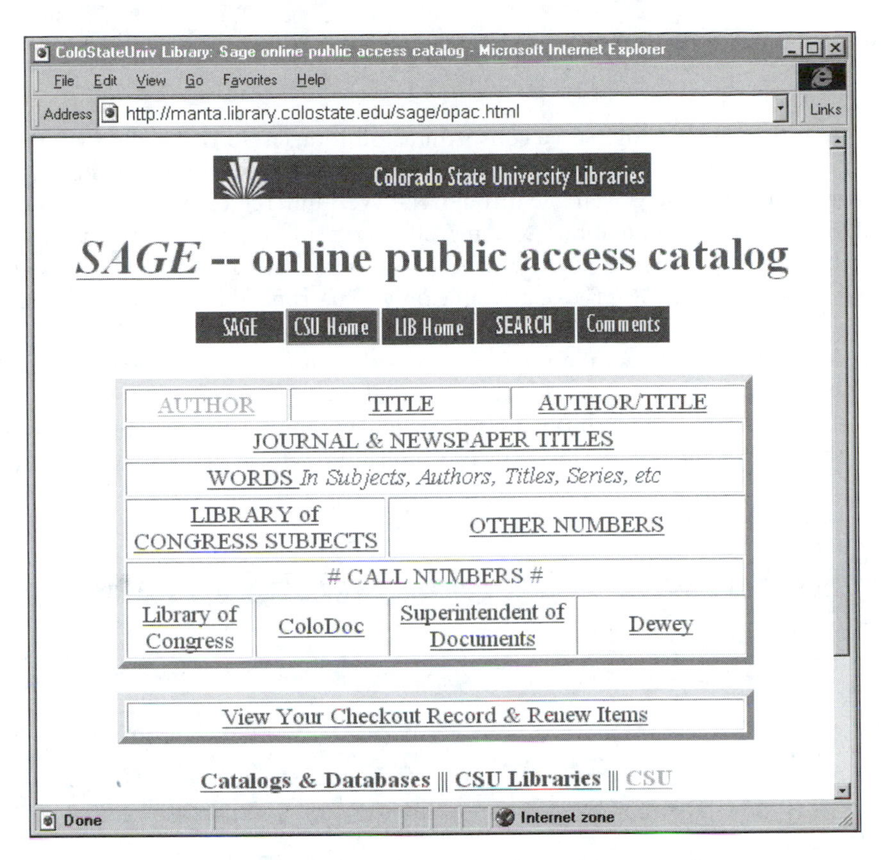

FIGURE 4.1 The SAGE Online Library Catalog—Web interface

```
Terminal - SAGE
File  Edit  Session  Options  Help
                         WELCOME TO SAGE
              Colorado State University Online Catalog

                   You may search by the following:

          A > Author
          T > Title
          B > Author/Title

          J > Journal & Newspaper Titles

          W > Words in Author, Title, Subjects, Series, etc.
          S > Library of Congress Subject Headings
          C > CALL Numbers/Other Numbers

          H > History of Searches Done
          V > View your Currently Checked out Items
          R > RETURN to previous menu

                  Choose one (A,T,B,J,W,S,C,H,V,R)
Connection established
```

FIGURE 4.2 The SAGE Online Library Catalog—Telnet interface

A few online catalogs serve consortiums of libraries. CARL—the Colorado Alliance of Regional Libraries—is the premier online catalog of this kind (see Figure 4.3). Via CARL, available through the Web and Telnet, you can access the holdings of 74 academic library systems, 20 magazine and journal databases, and hundreds of other online library catalogs. In addition to its public offerings, CARL offers a range of services available only to patrons of libraries who belong to the CARL consortium.

Although the primary reason writers consult online library catalogs is to find out what is available at a given library, writers also use library catalogs to find out what has recently been published on a particular topic. After searching several online library catalogs, writers can use interlibrary loan to order documents

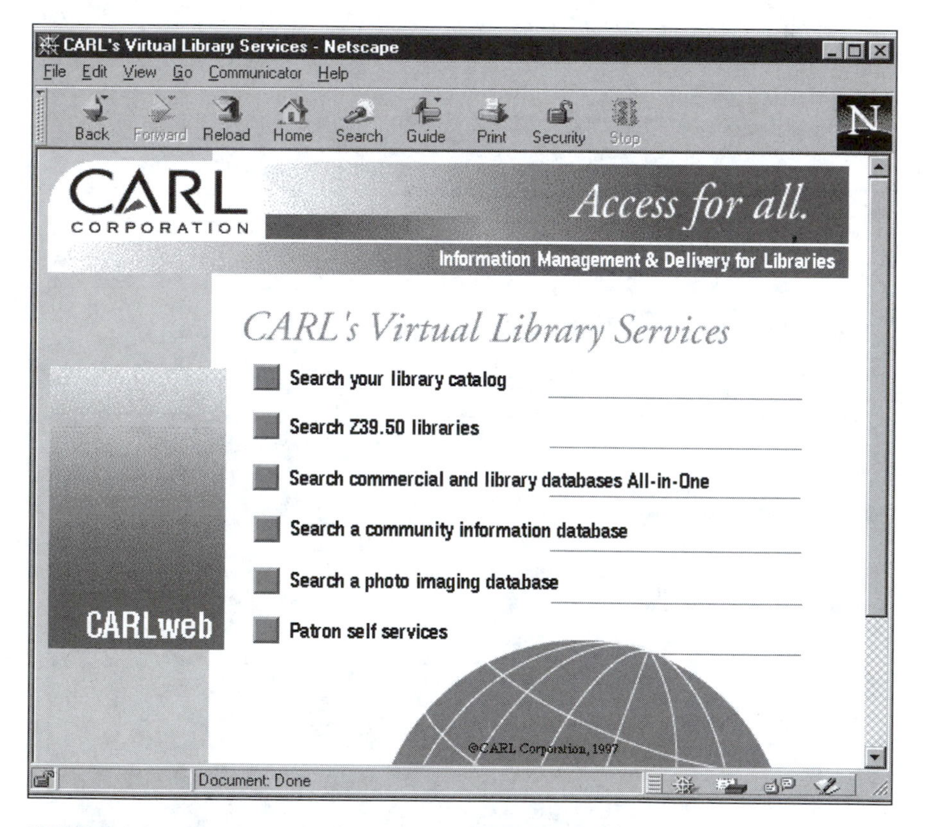

FIGURE 4.3 The CARL Online Library Catalog

that are not held by a local or institutional library. To learn the Web or Telnet addresses of online library catalogs, visit the CARL Website at <http://www.carl.org/lpart.html> or visit Yahoo's REFERENCE:LIBRARIES listings at <http://www.yahoo.com/Reference/Libraries>.

Databases

Databases provide information about specific subjects. Databases typically focus on either numeric or textual information. A database, for instance, might contain numeric information gained from the most recent United States Census. Most writers, however, are interested in textual databases, particularly those that focus on published documents.

Textual databases contain records, or individual entries on particular books, newspaper and journal articles, papers, Web sites, unpublished manuscripts, and so on. Each record is organized to support efficient searches of information. Database records typically include publication information, an abstract (brief summary) of the source, and key words that describe the content of the source. Some databases also provide access to the complete text of documents.

Thousands of databases are available on the Web, via Telnet, and on CD-ROM. Although some databases can be accessed without cost (you can access 20 online databases on CARL at <http://www.carl.org>), most databases are sold on a subscription basis. Libraries and corporations, for instance, receive online access to or CD-ROM copies of the databases to which they subscribe. Other databases are available only on a fee basis. These databases usually serve a specific discipline or profession, such as microbiology or contract law. You will be charged a fee to access these databases; you might also be charged a fee for a record or document that you download from the database.

Because of the high cost of many databases, the best place to use them is often your library. Libraries are investing an increasing amount of their acquisition budgets to provide their patrons access to databases. A result of the high cost of database subscriptions and fees is a tendency to subscribe to databases of interest to the largest number of patrons. Due to high costs, few libraries provide free access to specialized databases.

Examples of databases of interest to large numbers of patrons include:

- UnCover, a database of over 17,000 multidisciplinary journals. UnCover, available via the Web and Telnet, contains over seven million records of articles published since 1988 (Figure 4.4).
- ERIC, a database on educational issues that includes citations and abstracts of articles, conference presentations, manuscripts, and reports published since 1966. ERIC is available on CD-ROM and via the Web and Telnet through online library catalogs.

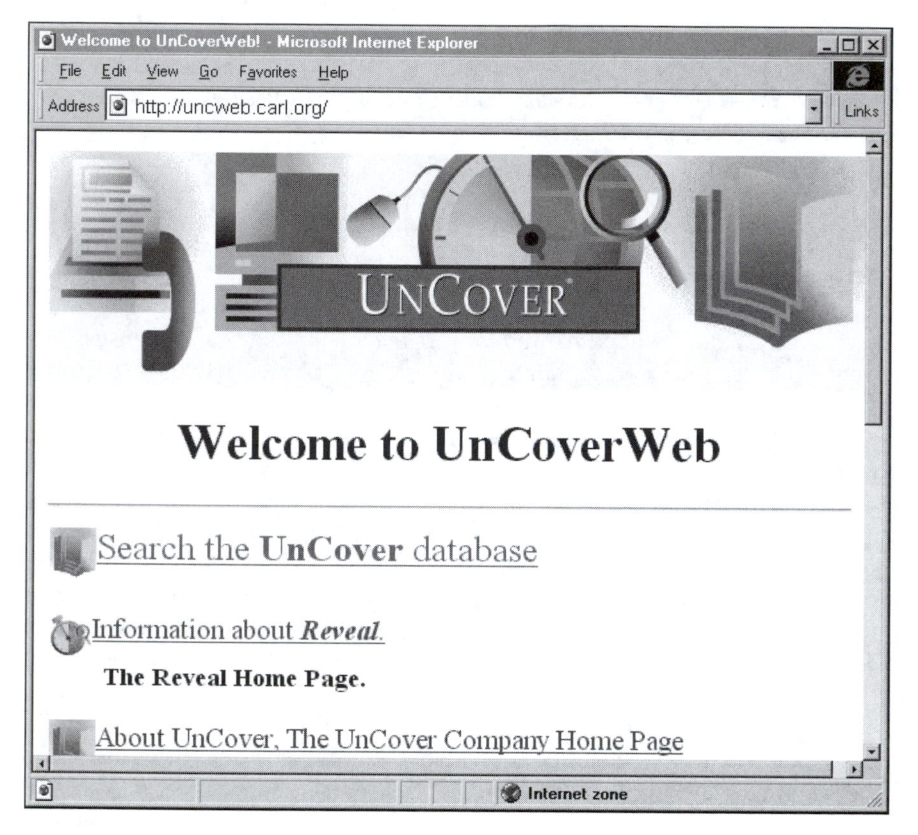

FIGURE 4.4 The UnCover database—Web site

- PsychINFO, a database of 1,300 publications in psychology and related fields available via the Web and Telnet, and on CD-ROM. Information is available on articles published since 1967.
- MLA Online, covering publications in literature, composition, linguistics, and foreign languages. MLA Online is available on CD-ROM.
- ABI/Inform, a business and management database containing more than 800,000 citations on business and management topics in over 1,000 publications world wide. Records date to 1971. Available on CD-ROM and via the Web and Telnet through online library catalogs.

Good places to locate databases include your local library and online library catalogs. You can also locate a large number of databases by conducting searches on Web sites such as Yahoo <http://www.yahoo.com>, Lycos <http://www.lycos.com>, and Excite <http://www.excite.com>.

Online Document Collections

Online document collections provide access to electronic versions of printed texts—typically, but not always, classic texts on which the copyright has expired. Most often, online documents are available in plain text. That is, they are not formatted in any way and can be read using any word processing program.

As the number and scope of online document collections grows, they are becoming increasingly useful to writers. By providing a service traditionally performed only by public libraries—providing access to books and articles—they are making it easier for writers to obtain information quickly and easily.

Some of the best online document collections include:

- Bibliomania: The Network Library <http://www.bibliomania.com>
- Complete Works of William Shakespeare <http://the-tech.mit.edu/Shakespeare/works.html>
- Electric Library <http://www.elibrary.com>
- Project Gutenberg <http://www.gutenberg.net>
- The B&R Samizdat Express <http://www.samizdat.com>
- The Etext Archives <http://www.etext.org>
- The New Bartleby: A National Digital Library <http://www.bartleby.com>
- The Online Books Page <http://www.cs.cmu.edu/books.html>

You can also find information about online document collections by searching for "electronic text," or "etext," on such search sites as Yahoo, Lycos, and Excite.

The World Wide Web

The World Wide Web, often referred to simply as the Web, is a subset of documents available on the Internet. Using the Web, you can easily view and move between documents, even when those documents are on computers literally thousands of miles apart. For example, you could view a document on a computer at the University of Minnesota in Minneapolis, then click on a link (a command containing information about the Internet location of a document) and view a document on a computer in Malaysia or New Zealand.

Web documents can appear as single pages or as sites containing multiple pages. You'll often run across the term *Home Page*, referring to the main page or introductory page in a Web site. Some sites contain only a few pages, while others contain thousands of pages. To visit a site, you need to know it's URL—uniform resource locator—which you can think of as a Web address (Panel 4.3).

Web documents can contain formatted and unformatted text, graphics, sound, video, links to other documents, and even programs that allow you to do such things as calculate how much money you'll need to save in order to ensure a comfortable retirement. Web documents are viewed through browsers such as

Panel 4.2
What Is Telnet?

Telnet is a software program that allows you to remotely control another computer on the Internet. To use Telnet, you need a Telnet program, access to the Internet from the computer you are using, and permission to use the remote computer. You'll also need to know the Internet name or address of the computer you want to access and the commands you will need to operate the remote computer's operating system and programs.

You can use Telnet to run programs on remote computers and to read information located on those computers. If you have an electronic mail account on a computer at your school or business, you might be able to use Telnet to read your mail from a remote computer. For example, as we were writing this book, Mike wanted to check his electronic mail while he was visiting relatives in Minnesota. He went to the local library, ran the Telnet program that was available on one of the computers connected to the Internet, logged into his account, and read his mail (Figure A). To make this possible he needed to know:

- If the computer in the library was connected to the Internet (it was)
- If the computer had a Telnet program (it did)
- The name of the computer he wanted to connect to (in this case, lamar.colostate.edu)
- His login name and password (which we won't disclose here)
- The commands needed to run the operating system on the remote computer (in this case, UNIX commands)

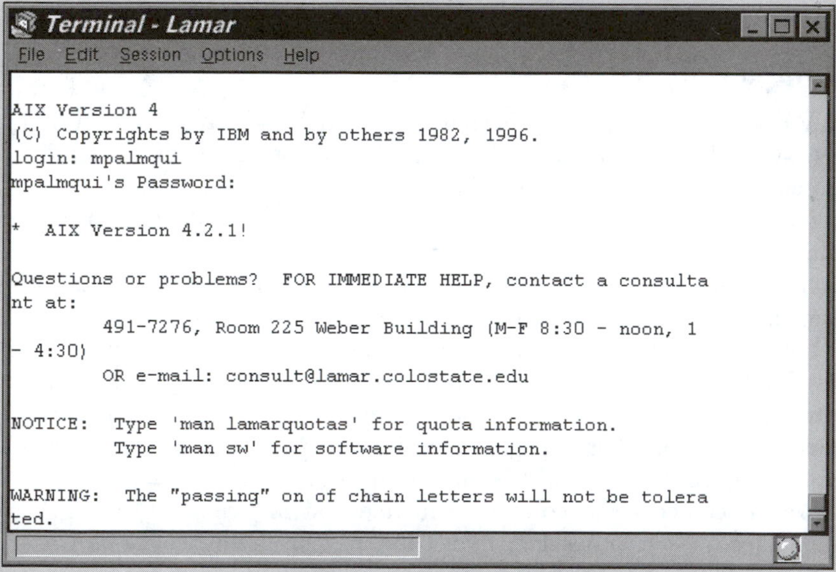

Terminal - Lamar

File Edit Session Options Help

```
AIX Version 4
(C) Copyrights by IBM and by others 1982, 1996.
login: mpalmqui
mpalmqui's Password:

*  AIX Version 4.2.1!

Questions or problems?  FOR IMMEDIATE HELP, contact a consulta
nt at:
        491-7276, Room 225 Weber Building (M-F 8:30 - noon, 1
- 4:30)
        OR e-mail: consult@lamar.colostate.edu

NOTICE:  Type 'man lamarquotas' for quota information.
         Type 'man sw' for software information.

WARNING:  The "passing" on of chain letters will not be tolera
ted.
```

FIGURE A Connecting to another computer using Telnet

Panel 4.3
What Is a URL?

The key to reading files on other computers is the URL (Uniform Resource Locator). A URL contains information that directs your browser to a particular file in a particular directory on a particular computer somewhere on the Internet. As an example, the URL, <http://www.colostate.edu/Depts/WritingCenter/index.html>, directs you to the home page of the Online Writing Center at Colorado State University. Each part of the URL has a purpose:

- "http://" tells the browser that the file can be accessed using the Hypertext Transport Protocol. In other words, it tells the browser that it will be reading a Web page.
- "www.colostate.edu" tells the browser which computer to look up—in this case, the main Web server at Colorado State University. Computer names can be deciphered by looking at the information separated by periods. In this case "www" is the name of the specific computer that you connect to at Colorado State University. "colostate" is the name of the institution that owns the computer. And "edu" indicates that the institution is a member of the educational domain. Some other domains include ".com," or commercial, ".gov," or government, and ".org," or noncommercial organization.
- "/Depts/WritingCenter/" tells the browser to look in the "/Depts/WritingCenter" directory on the Web server
- "index.html" indicates which file to read in the directory

Netscape Navigator and Microsoft Internet Explorer. Browsers interpret Web documents and present them on your computer. Browsers are much like word processing programs. They allow you to read files, save those files on your computer, and copy text and other elements on a page. Unlike word processing programs, they also allow you to use your modem or computer network to connect to and read files on other computers.

Since its introduction in 1992, the Web has grown phenomenally. There are millions of Web sites containing many more millions of pages. In 1996, it was widely reported that, in large part due to the Web, the amount of text available online exceeded the amount of text available in print form. Recent expectations are that over 200 million people will be using the Web by the year 2000.

Writers can turn to the Web for information on almost any subject. Web sites contain information about individuals, institutions, corporations, organizations, and government agencies. Some Web sites are developed as public services. Others are developed to promote causes or advertise products and services. Still others are developed to entertain readers and, often, the site developers.

Unlike materials in a library, the quality and accuracy of documents on the Web varies greatly. You can turn to Web sites sponsored by the Library of

Congress or the Smithsonian with a great deal of confidence about quality and accuracy. But it's unclear, without evaluation, whether a hypothetical Web site called "Stephen's Little Shop of Physics" contains useful information. You'll be more confident of the value of the site if it reflects the latest thinking of Stephen Hawking, considered by many the world's leading theoretical physicist, and less confident if it reflects seventh-grader Stephen Smith's thinking on the subject.

This book has directed your attention to Web sites on a range of topics. These sites contain accurate and useful information. But you'll encounter many more sites as you search the Web. The last section of this chapter provides an overview of evaluation techniques for Web sites and other online information sources.

Gopher

Gopher is a text-based predecessor to the World Wide Web. Designed to support easy access to information found on the Internet, Gopher provides a simple interface that takes the place of tools such as Telnet and FTP. To locate information on a Gopher site, you follow menus to documents you would like to read (Figures 4.5 and 4.6).

Gopher was developed at the University of Minnesota and bears the name of the University's mascot. Gopher also stands for "go-for," the underlying idea of a Gopher server. Several Gopher client programs are available to view texts on a Gopher site. Web browsers can also read texts on a Gopher site.

Before the Web became popular, many universities and government agencies published a great deal of information on Gopher sites. Although much of

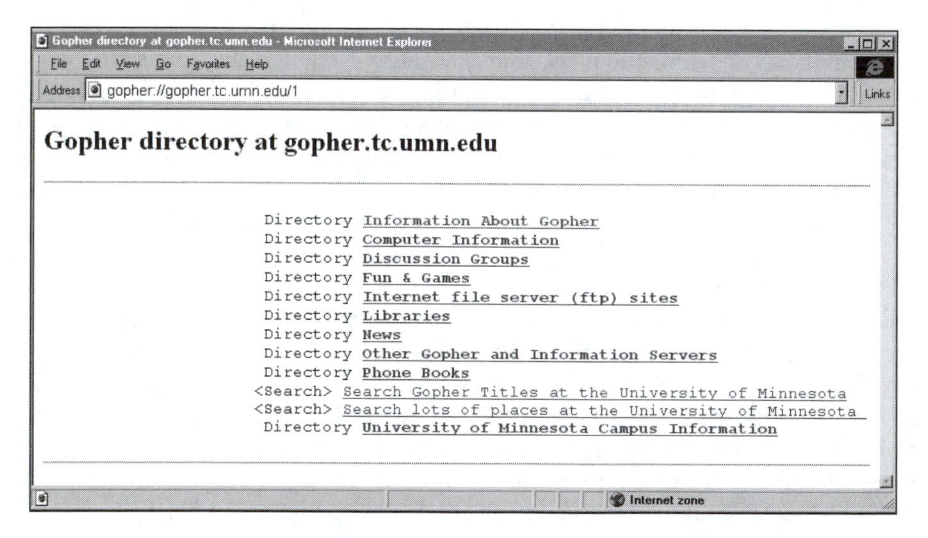

FIGURE 4.5 The main Gopher site at the University of Minnesota—Web view

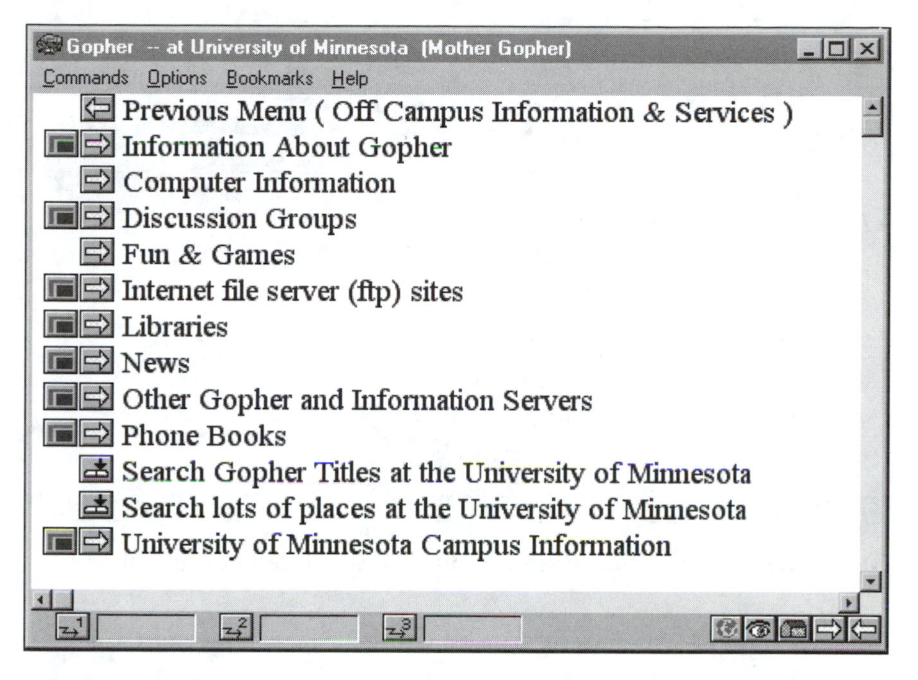

FIGURE 4.6 The main Gopher site at the University of Minnesota—Gopher client view

that information has since been converted into Web format, some materials continue to be available only through Gopher.

Newsgroups

Newsgroups are Internet discussion forums on specific topics. There are tens of thousands of newsgroups on the Internet, addressing topics ranging from athletics to zoology. You can read messages posted to a newsgroup using a newsgroup reader. A newsgroup reader functions much like an electronic mail program. You can read messages posted to newsgroups, send messages to the newsgroup, and send electronic mail to individuals who posted to the newsgroup. Popular Web browsers such as Netscape Navigator and Microsoft Internet Explorer include built-in newsgroup readers (Figure 4.7). You can also obtain free newsgroup readers on the Internet. In addition, several commercial newsgroup readers are available. If you read newsgroups a great deal, the additional features provided by commercial newsgroup readers more than justify their cost.

Newsgroups use a technique called threading to organize messages on particular topics (Figure 4.8). Threading allows you to identify threads that interest

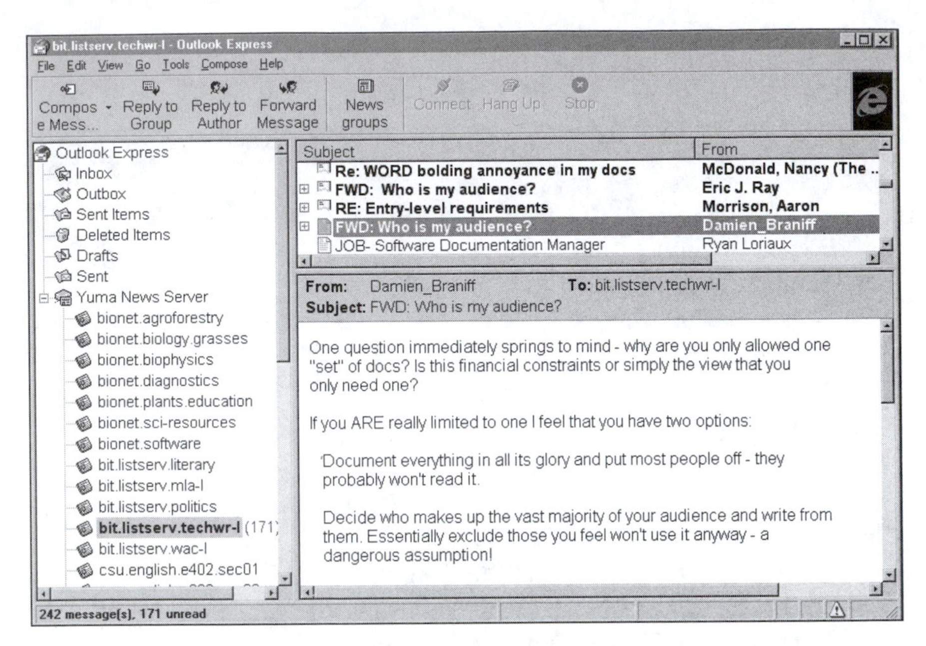

FIGURE 4.7 The built-in newsreader in Microsoft Internet Explorer

Re: Disney Politics (was Re: Disney Marathon - rec.running - Netscape Discussion

File Edit View Go Message Communicator Help

Get Msg New Msg Reply Forward File Next Print Security Mark Stop

rec.running Total messages: 643 Unread messages: 627

Subject	Sender	Date	Priority
Re: Do elite runners drink durin...	Paul French	Mon 12:39	
Re: Disney Marathon	Larry	Sat 12:19	
Re: Disney Marathon	Mike Tennent	Sat 2:18	
Re: Disney Marathon	Rob Madsen	Sat 17:29	
Re: Disney Marathon	Doug Freese	Sat 15:47	
Disney Politics (was Re: ...	Sam Callan	10/23/97 11:56	
Re: Disney Politics (w...	Dean Ouellette	Mon 10:23	
Re: Disney Politic...	Brian J. Zimbelman	Mon 12:53	
Re: Disney Politic...	Doug Freese	Mon 15:37	
Re: Disney Politic...	frodojrr	Mon 21:15	
Re: Disney Politic...	Sam Callan	Tue 19:15	
Re: Disney Marathon	Larry	Mon 14:43	

Hide: Disney Politics (was Re: Disney Marathon

Subject: Re: Disney Politics (was Re: Disney Marathon
Date: Wed, 29 Oct 1997 02:15:26 GMT
From: marathonman
Organization: MindSpring Enterprises
Newsgroups: rec.running
References: 1 , 2 , 3 , 4 , 5

Document: Done

FIGURE 4.8 A threaded discussion in rec.running

you and to ignore those that do not. Since thousands of people read newsgroups, you'll often find that they contain a great deal of information that you may not need.

Some, but by no means all newsgroups are archived. Archives contain posts made to the newsgroup over an extended period of time, ideally from its inception to the present. Many archives are searchable; and all, by their nature, are browsable. If you cannot find an archive for a particular newsgroup, conduct a search on DejaNews <http://www.dejanews.com> for posts to that newsgroup. A search on DejaNews for the newsgroup rec.running, for example, yielded nearly 3,000 posts to the newsgroup.

In addition to reading newsgroups to locate information, you can also post messages to a newsgroup requesting information from other readers. Use this technique if you have sufficient time to wait for a response from other readers.

To locate newsgroups, consult:

- Deja News, a searchable and browsable list of newsgroups and newsgroup posts <http://www.dejanews.com>
- Yahoo's directory of newsgroups and newsgroups postings <http://www.yahoo.com/News_and_Media/Usenet/Newsgroup_Listings>

Mailing Lists

Mailing lists allow individuals to send electronic mail messages to a list of people interested in a particular topic. Like newsgroups, they support online discussion of a particular topic, such as writing. Mailing lists are maintained by software, such as Listserv, that distributes and archives messages sent to members of a group. The mailing list sponsored by the Alliance for Computers and Writing, ACW-L, for instance, focuses on issues related to the use of computers for writing and writing instruction.

Before you can send a message to a mailing list, you must subscribe to the list. This involves sending an electronic mail message to the computer—or the list server—that runs the mailing list. The content of the message varies according to the type of software used to run the mailing. You can usually find out what message to send, however, by consulting a Web site such as Liszt, a searchable directory that contains information about more than 80,000 public and private electronic mail discussion lists <http://www.liszt.com>, or Catalist, a searchable directory of over 17,000 public electronic mail discussion lists <http://www.lsoft.com/catalist.html>. Once you have subscribed to a mailing list, messages posted to the mailing list are sent to your electronic mail account.

One of the drawbacks to mailing lists, as you might imagine, is the number of messages you can receive from them in a given day. If you subscribe to an active mailing list, be prepared to deal with as many as 100 messages each day. If you subscribe to several mailing lists, you may find yourself inundated with electronic mail messages.

To deal with the large number of messages generated by a mailing list, you can tell the computer that runs the mailing list to digest your messages. The digest command tells the mailing list to send multiple messages in a single file.

You can use mailing lists for the same purposes that you use newsgroups. You can search a mailing list's archives, if one is available. You can also post messages requesting information from other readers.

Chat Channels and MOOs

Chat channels (or rooms) and MOOs (Multi-User Dungeons—Object Oriented) share many similarities. Both allow people to type messages to each other in real time, even though they might be separated by thousands of miles. The messages appear on the screen in the order in which they are written. As you might imagine, if there are several people (or several dozen) chatting at the same time, it can get a bit confusing. Confusing or not, many writers find the dynamic nature of Chat and MOOs stimulating.

The fundamental differences between Chat and MOOs arise from their different origins. Chat is based, at least historically, on a UNIX utility called Talk. More immediate than electronic mail, Talk allows people to quickly ask each other questions or to engage in extended conversations. Like electronic mail, Talk evolved to serve social as well as utilitarian purposes. Eventually, Internet Relay Chat was developed allowing people to communicate using Chat via the Internet. Similar systems were developed for commercial service providers, such as Prodigy and CompuServe. Users of Internet Relay Chat connect to IRC servers and join discussions taking place on various channels. IRC servers communicate with each other, so users connected to one channel on a particular server can communicate with users connected to that channel on other servers.

Like Chat channels, people who enter a MOO can talk with one another. There can also be many people in a particular room. However, MOOs have several additional capabilities. The most important difference is that a MOO uses textual descriptions to define the space in which you interact with others. The notion of a room in a MOO is quite real—you move from room to room, down hallways, into courtyards, and so on. Each time you move into a new space, you can learn who is in the room, what they look like, what they are doing, and what the room looks like. In addition, you may find yourself talking to a robot rather than a person. Robots are programs that are created by the designer of a MOO. Sometimes these robots are so well designed that you may confuse them with real people. People who use MOOs can also create descriptions of themselves so that other people can "see" them when they enter a room in a MOO. Using the @DESCRIBE command, you can create a character that others will see when you enter a room. Your character can have a name, a gender, and a brief text description—none of which may have anything to do with who you are when you aren't online.

The Origins of MOOs. Multi-User Dungeons (MUDs) emerged in the 1980s when devotees of the game Dungeons and Dragons began searching for ways to play their popular game online. In Dungeons and Dragons, players attempt to find their way through the dangers of a dungeon that has been designed by a "Dungeon Master." Dungeon Masters designed their dungeons in deliberately confusing and dangerous ways, with clear rules about what happens when a particular playing piece enters each room or hallway in a dungeon. The game is well suited to computer-based play and, over time, computer-based Dungeons and Dragons games appeared. In the late 1980s, a newer version of MUDs developed—MOOs, (Multi-User Dungeons—Object Oriented). Object oriented programming allows designers and players to more easily customize the dungeons and the appearance of the players.

Uses for MOOs. What does this have to do with writing? As they have with other online technologies, writers have learned to use Chat and MOOs in ways that were not anticipated by the inventors of the programs. Among other things, Chat and MOOs allow writers to share ideas, to collaborate on projects, and to ask for information about particular topics. Many writing classes use Chat and MOOs to carry out collaborative brainstorming sessions, to critique drafts of papers, and to engage in role-playing games such as Devil's Advocate. In general, Chat and MOOs provide writers with the opportunity to talk about their writing with other writers.

To use Chat and MOOs, you need to obtain appropriate software. To use Chat, you should obtain an Internet Relay Chat client. These are available as part of many popular Web browsers, such as Netscape Navigator and Microsoft Internet Explorer. Other "freeware" IRC clients can be obtained at sites listed under Yahoo's COMPUTERS AND INTERNET: INTERNET: CHAT: IRC: SOFTWARE: CLIENTS AND SERVERS. You can also obtain information about IRC clients by searching for "IRC" on a Web search site such as Lycos <http://www.lycos.com> or Excite <http://www.excite.com>.

To use a MOO, you need a Telnet program or a Telnet program that has been adapted for use on a MOO (usually by adding a small window where you can type your comments). Most operating systems, such as Windows 98 or the Macintosh OS, include a simple Telnet program. You can obtain more sophisticated versions of Telnet at a number of Web sites. Yahoo provides a listing of sites at COMPUTERS AND INTERNET: SOFTWARE: INTERNET: TELNET. You can also obtain information about Telnet-related sites by searching for "Telnet" on a Web search site. To obtain information about Telnet programs that have been adapted for use on a MOO, visit one of the following:

- Yahoo's Recreation:Games:Internet Games:MUDs, MUSHes, MOOs, etc.:MOOs
- The Netoric Project's General MOO/MUD Information Page <http://bsuvc.bsu.edu/~00gjsiering/netoric/moo.html>
- MU* Clients <http://homepages. together.net/~shae/client.html>
- Chris' MOO Page <http://www.cms.dmu.ac.uk/~cph/moos.html>
- The Internet Public Library's MOO Resources <http://www.ipl.org/moo/staffinfo.html>

To find general information about Chat rooms and Internet relay Chat, check out:

- Yahoo's Computers and Internet:Internet:Chat
- Internet Relay Chat: What You Really Need to Know About IRC <http://www.cire.com/patrick/irc.html>
- Efnet #IRChelp help archives <http://www.irchelp.org>
- A Short IRC Primer <http://www.irchelp.org/irchelp/ircprimer.html>

Some, but by no means all MOOs maintain archives. The Netoric Tuesday Café, for instance, a group of writers and writing teachers who meet in a MOO on a weekly basis, maintain archives that can be requested from the Netoric Project's Web site <http://bsuvc.bsu.edu/~00gjsiering/netoric/logs.html>. Chat channels seldom maintain archives. However, you can save transcripts of any Chat or MOO discussion in which you are involved. The commands for saving a transcript depend on the program you are using to participate in an online discussion.

SEARCHING AND BROWSING ONLINE INFORMATION

Given the wealth of information available online, it can be surprisingly difficult to locate precisely what you're looking for. You can benefit, however, from strategies that help you search for and browse potentially useful information.

Browsing and searching are terms borrowed from traditional library research. Browsing—wandering through the stacks of a library and looking at books that happen to be near at hand—is anything but systematic. Yet it's part of the systematic approach taken by most library researchers. Experienced researchers know that wandering through an area devoted to a particular subject can help them locate books and articles that they might not have found had they relied only on the library's card catalog or subject indexes. Browsing is a process that relies on serendipity, but it has proven time and again to be an effective tool for locating valuable information on a topic.

Browsing has taken on an expanded meaning since the emergence of the World Wide Web. The very nature of the Web supports browsing. Just as it's easy to pick up and glance through a book next to one you had found through a catalog or subject index, readers of Web documents find it easy to click on a link and move to a new page. The practice of following links from document to document—or browsing the Web—has become a modern commonplace. Some writers follow links with abandon, often moving so far afield from the document they'd originally wanted to read that they find it impossible to find their way back.

Searching stands in direct contrast to browsing. Searches are systematic attempts to locate relevant information on a particular subject. In libraries, searches are conducted using subject indexes or online card catalogs. On the Web, searches are conducted using search engines or information directories. Yet searches can be far from precise. On the Web, searches can sometimes yield lists containing thousands of documents that might be relevant to the topic. The operative phrase, however, is "might be relevant." Unless a search is carefully planned, you can end up with a haphazard array of documents that have little in common with each other, let alone with your research interests.

In this section, we discuss techniques you can use to locate information online, including searching online texts, searching database fields, using wildcards and exact phrases in your searches, conducting Boolean searches, limiting searches by publication information, searching Web indexes and directories, searching "search sites" on the Web, browsing the Web systematically, searching GopherSpace, and consulting newsgroups and mailing lists.

Each of these techniques involves searching or browsing for information. As you read about each technique, consider how well it fits your approach to locating information.

Searching Documents and the Full Text of Database Records

Searching documents and the full text of database records involves looking for any occurrence of a word or phrase in one or more documents or database records. Unlike searching database fields, discussed in the next section, this kind of search looks for occurrences of words or phrases in any part of the document or database records.

The terms you search for are typically referred to as keywords. In general, keywords are used to identify the subject of a search. A search for information related to writing with computers, for instance, might use the keywords "writing" and "computers." If you are looking for information about specific processes writers use on a computer, such as brainstorming or collecting information, you might include additional keywords, such as "brainstorming," "generating ideas," or "searching."

When the search is conducted, documents and database records are usually ranked according to how many of the keywords appear in each document or record and how often each term appears.

tip 4.1 ■ Keep Lists of Keyword Searches

As more and more information becomes accessible online, you're likely to find yourself spending more of your time searching for information. One of the keys to effective searching is to use multiple keywords in varying combinations. We've noticed, however, as have our students and colleagues, that we sometimes forget which words and which combinations we've already used.

You can increase the effectiveness of your searches—and reduce the time it takes to conduct them—by keeping a list of the words you've used for each search. As you complete each search, jot down or list in your word processing program the number of hits (records) that the search yields and indicate the percentage of sources that were actually relevant to your needs.

Records of your search terms can come in handy not only while you make your initial searches, but also later on, should you choose to search again on the same topic.

Searching Database Fields

In addition to searching for keywords anywhere within documents or records, you can conduct focused searches for keywords within a specific field or fields in a database. Databases use fields to simplify the process of searching for information. You can search for occurrences of specific keywords in document titles or abstracts, for example, or you can search for specific authors in the author field.

Many databases use fields to identify documents according to subject. Sometimes these fields are called descriptor or subject fields. If you know the specific keywords associated with a subject, you can search the descriptor or subject fields for database records on a particular subject. Some databases, such as ERIC, use the phrase *major descriptor* to refer to larger categories of information, and the phrase *minor descriptor* to refer to subcategories of information.

Most databases that use descriptor or subject fields include a list of keywords related to the subject. You can also identify subject keywords by doing a general search and viewing the descriptors listed in particular records. In ERIC, a search on "computers" and "writing" yielded a list of 1,570 records containing the two keywords—certainly too many to comfortably search in an afternoon. One promising title, Martha Pennington's "Writing the Natural Way: On Computers," contained the major descriptors "writing skills" and "word processing." A second search, restricted to the two major descriptors, yielded a more manageable list of 68 records (Figures 4.9 and 4.10).

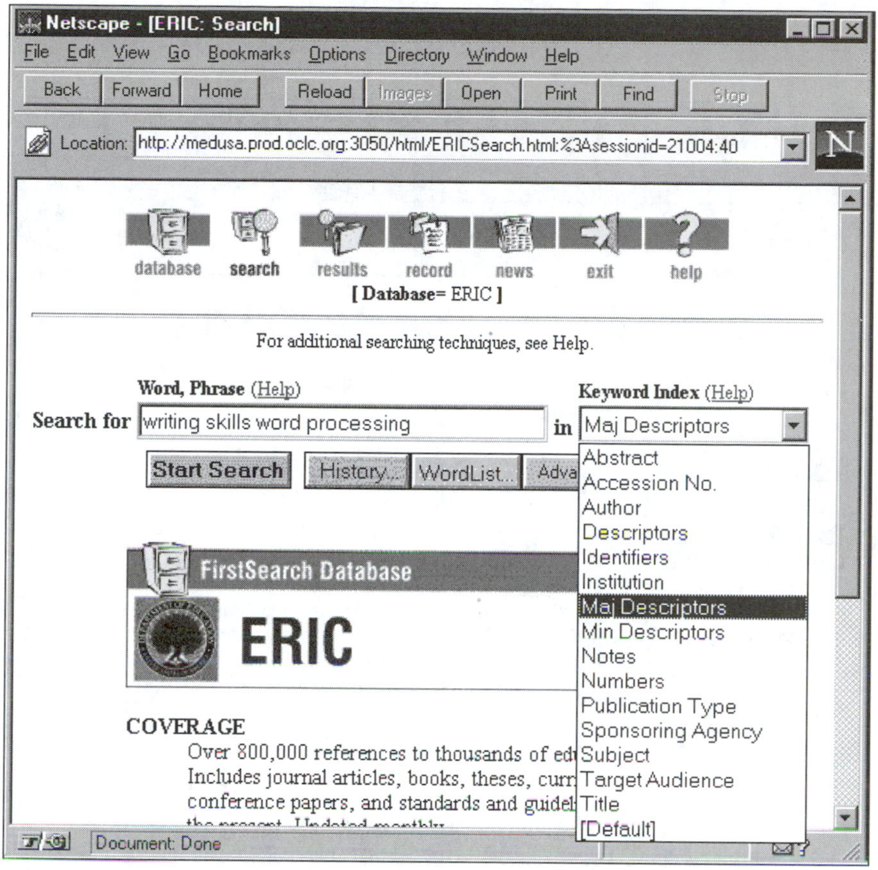

FIGURE 4.9 Searching the ERIC database using major descriptors

tip 4.2 ■ Save Search Results to a File

A useful technique for keeping track of your search terms and results is saving the results of your search to a file. Many CD-ROM databases and a growing number of databases on the Internet allow you to save the records you retrieve during a search. Often, you can mark individual records and save those for later analysis. In some databases, you can also define which parts of a record to save. For instance, you might choose to save only publishing information and discard any abstracts or annotations.

If you're doing research in a library, it's a good idea to bring one or more floppy diskettes with you. They'll come in handy if you find yourself using a database that allows you to save the results of your search to a diskette.

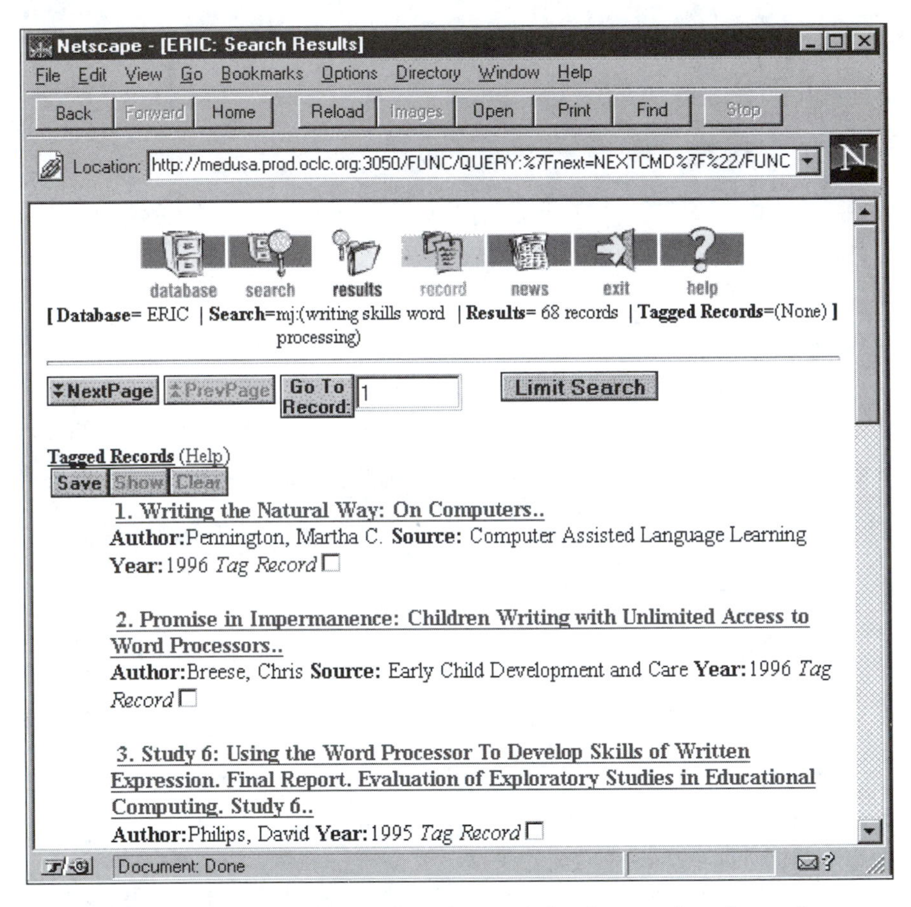

FIGURE 4.10 Results of searching the ERIC database using the major descriptors "Writing Skills" and "Word Processing"

Searching with Wildcards

The best things about computers is that they do *exactly* what you tell them to do. The worst thing is that they do *only* what you tell them to do. If you conduct a keyword search for the terms "computers" and "writing," you'll find every source the computer can locate that contains those two terms. However, the computer won't tell you about sources that contain the terms "computers" and "writer," or "computers" and "write."

Wildcards allow you to search for words that are similar in form. For instance, if you want to search the AltaVista Website <http://www.altavista.digital.com>

for all records containing the keywords "writer," "writing," "write," "written," and "writes," you can use a wildcard to search for all of these terms at once (Figure 4.11). Typical wildcard symbols include an asterisk (*) to denote any combinations of letters or symbols, and a question mark (?) to denote single characters. A wildcard search for computers and writer, writing, write, written, or writes would look like this: computers writ*.

Many search engines on the Web and in databases do not support wildcard searches. Consult the online help to learn whether you can use wildcards in your searches and, if so, the specific symbols used by the search engine.

Searching for Exact Phrases

When a search engine looks for occurrences of keywords in a document or database record, it usually looks for occurrences of each keyword regardless of where it appears in relation to the other keywords. For instance, a search conducted on

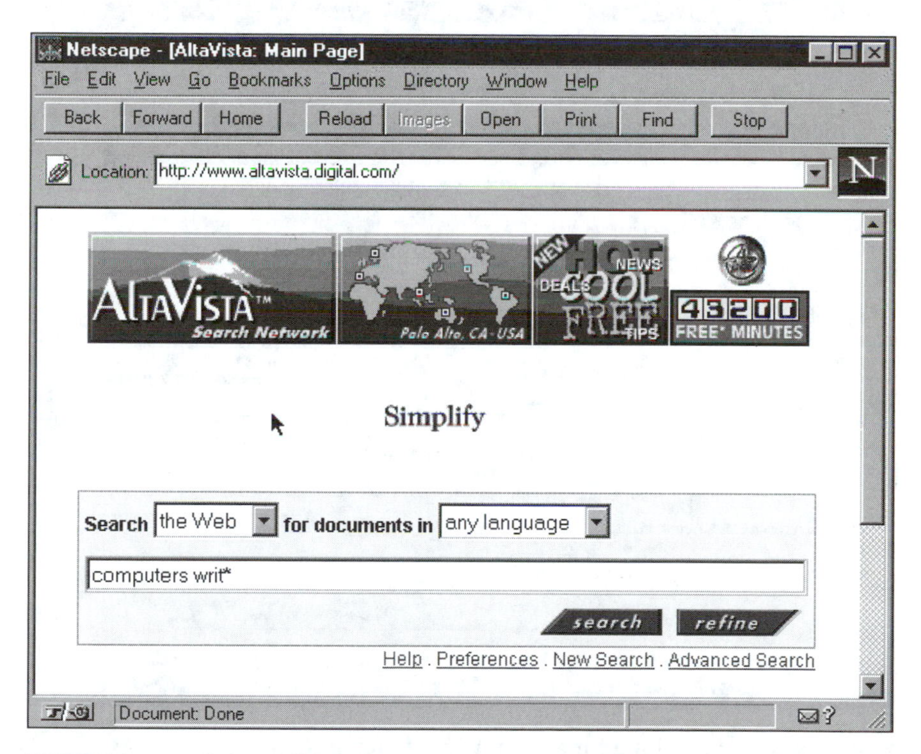

FIGURE 4.11 Using wildcard symbols to search AltaVista

the keywords "writing" and "process" would provide a list of documents in which one or both keywords appeared. Some of those documents might contain the phrase "writing process," but others might not. Some documents, for instance, might provide directions for writing reports about corn drying processes—hardly what you would be looking for.

To increase the likelihood that you will find what you are looking for, many search engines allow you to search for exact phrases. The specifics of searching for exact phrases vary from search engine to search engine. A common technique for searching for a phrase, however, is to enclose the phrase in quotation marks. Another is to use a Web form to enter the phrase (Figure 4.12).

Searching for exact phrases, such as "writing process," allows you to conduct more refined searches. An advantage—particularly when you're working on the Web—is that your search is likely to exclude a large number of irrelevant documents or database records that include the keywords you searched for but otherwise have nothing to do with the subject of your search.

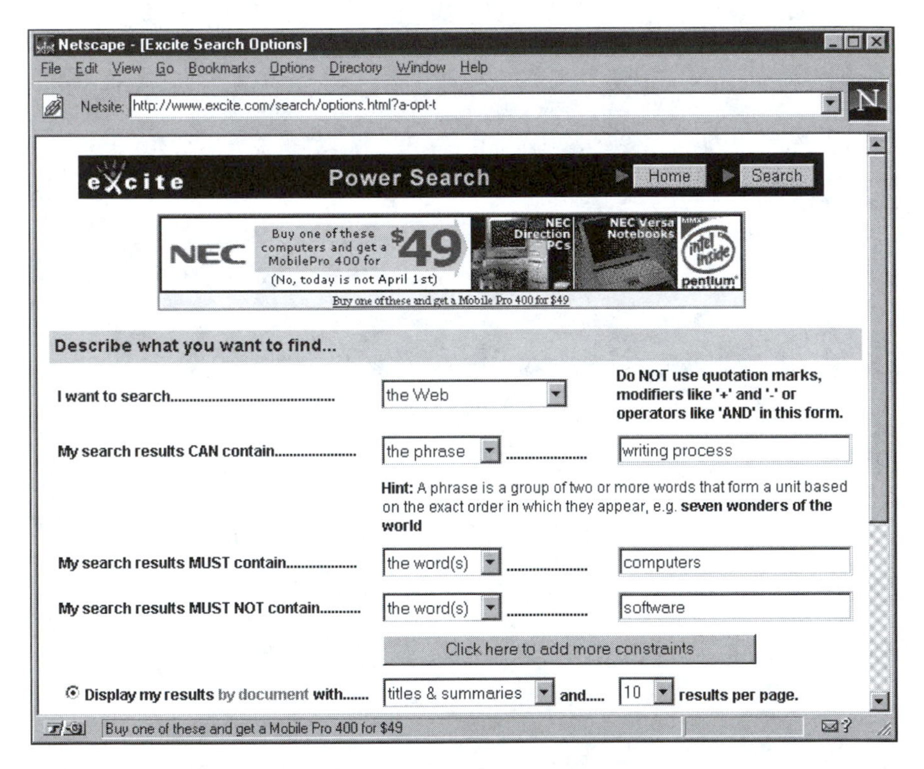

FIGURE 4.12 Searching for an exact phrase using Excite

Be aware, however, that searching for exact phrases can exclude relevant documents or database records from your search results. Searching for the phrase "writing process" will not let you know about documents containing phrases such as "writing is a process" and "the process of writing."

Using Boolean Terms in a Search

The most powerful and flexible way to search is a technique called Boolean search. Named for nineteenth century mathematician George Boole, Boolean search is used to precisely define the relationships among keywords in a search.

Boolean search uses three key operators—AND, OR, and NOT—to define a search. By convention, the three operators are capitalized, although it is not usually necessary to do so when using them in a search. Table 4.1 illustrates how Boolean operators function in a search.

By default, most search engines on the World Wide Web use the Boolean OR when they conduct searches for keywords. That is, they search for documents or database records that contain one or more of the keywords. Using OR will result in the largest set of search results. In contrast, many of the search engines used in databases use the Boolean AND when they search for keywords. That is, they search for documents or database records containing all of your keywords. Using AND results in a smaller set of search results. Using NOT reduces your results even further.

Some search engines support an expanded set of Boolean operators. Operators such as ADJ, BEFORE, NEAR, and FAR can be used in combination with or in place of AND, OR, and NOT:

- ADJ specifies that keywords must be adjacent to each other. For instance, you could search for all documents or database records containing two keywords in any order, such as "writing process" or "process writing."

TABLE 4.1 How to use Boolean search strategies

Search:	Result: All Documents or Database Records Containing
computers AND writing	both of the keywords
computers OR writing	either or both of the keywords
computers NOT writing	the keyword "computers" but not the keyword "writing"
computers AND writing NOT software	the keywords "computers" and "writing" but not the keyword "software"
computers OR writing NOT software	the keyword "computers" or the keyword "writing" but not the keyword "software"

- BEFORE specifies that keywords must appear in a particular order—one keyword must come before the other. Used in combination with ADJ, you can specify a search for exact phrases, such as "technical writing."
- NEAR specifies that keywords must be within a certain distance of each other. The advanced search engine on Lycos <http://www.lycos.com>, for instance, defines NEAR as within 25 words of each other. In some search engines, including Lycos, you can override the default distance by specifying the precise number of words you want to set as the limit.
- FAR, in contrast, specifies that words must be a certain distance apart. The advanced search engine on Lycos defines FAR as more than 25 words apart, but you can specify your own limit.

In addition, many search engines support the use of parentheses to specify complex relationships among keywords. For instance, the search *writing AND computers NOT (software OR marketing OR technical)* will search for all documents or database records containing the keywords "writing" and "computers" but excluding any that contain the keywords "software," "marketing," or "technical."

Search engines that support Boolean search allow you to create highly specific searches, increasing the likelihood that the results of the search will be relevant to your needs.

Limiting Searches by Publication Information

Many search engines allow you to restrict your search of documents to those containing specific types of publication information (Figure 4.13). You can limit your search to documents:

- Written by specific authors
- Published on, before, or after a particular date
- Of particular types

A database, for instance, might allow you to search for all documents published in 1997 by the author Kate Kiefer or for newspaper articles published after 1993. Similarly, many Web search engines allow you to limit your search to Web documents or Usenet groups. These limits can be used in place of or in addition to other kinds of searches, such as wildcard searches, Boolean searches, or searches for exact phrases.

Searching Web Indexes

A Web index is a database that contains information about other Web sites. Lycos, one of the best known Web search sites, is an index. It contains millions of records that you can search in highly sophisticated ways. Each record contains information about the title, location, and content of a Web site or Web page.

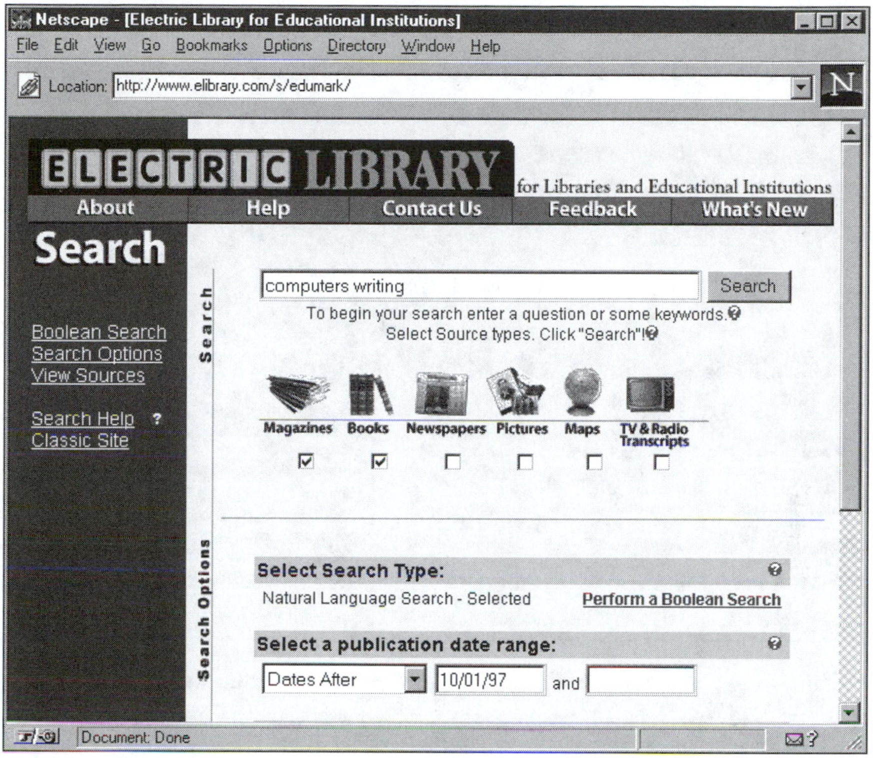

FIGURE 4.13 Limiting a search by publication information

The best Web indexes provide both simple search engines (essentially, keyword searches using Boolean AND) and more sophisticated search engines. Simple search engines are usually displayed prominently on the home page of the search index. Additional information about how to use the search engine, as well as a link to a more powerful search engine, are usually located near the simple search engine (Figure 4.14).

Leading Web Indexes include:

- AltaVista <http://www.altavista.digital.com>
- Excite <http://www.excite.com>
- HotBot <http://www.hotbot.com>
- Infoseek <http://www.infoseek.com>
- Lycos <http://www.infoseek.com>
- AOL Net Find <http://www.aol.com/netfind/>

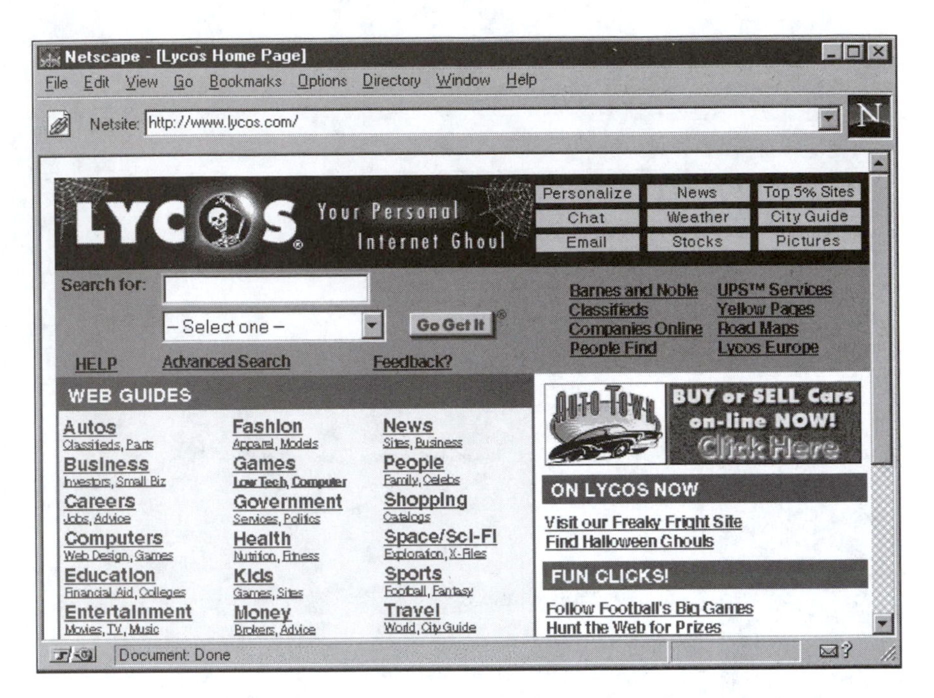

FIGURE 4.14 Access to Help and Advanced Search on Lycos

Despite differences in their overall emphasis, Web indexes and directories (discussed in the next section) have grown increasingly similar over the past few years. Although a given site will focus on providing either an index or a directory, most support both types of search.

Searching Web Directories

Like Web indexes, Web directories provide information about the content of Web pages. In addition, most provide search engines that rival the complexity of those found on Web indexes. If you used only their search engines, in fact, you'd notice little difference between Web directories and Web indexes.

The distinguishing feature between Web indexes and Web directories is the latter's organization of information into directories and subdirectories. Like the folders on your computer's desktop (see Chapter 7), directories and subdirectories contain information of increasing specificity. In the same way that you navi-

Panel 4.4
What Is HTML?

Web documents are formatted using Hypertext Markup Language (HTML), a subset of the document description language Standardized General Markup Language (SGML). HTML consists of a set of tags that tell your browser how to display a particular file. These tags are contained in plain text files, allowing computers of all kinds to read the same files. Tags tell a browser how to display text or other elements in an HTML file—they are not themselves displayed in a browser. Many tags deal with formatting issues, while others deal with setting up links to other documents. For instance, the tag **<b>** tells the browser to display text following the **<b>** in boldface. The tag **</b>** tells the browser to stop displaying text in boldface. Other common tags include:

- **<i>** italic face
- **<u>** underline
- **
** insert line break
- **<p>** insert paragraph break
- **<hr>** insert horizontal line

The following passage, formatted in HTML, appears like this when viewed as plain text:

<hr><p>

HTML tags tell your browser to display text in various ways, including **<b>**bold text**</b>**, **<i>**italicized text**</i>** and **<u>**underlined text**</u>**.

<p>

You can also use HTML tags to display text using **<b><i><u>**combinations**</u></i></b>** of these styles.**<p>**

<hr>

The passage appears like this when viewed in a browser:

HTML tags tell your browser to display text in various ways, including **bold text,** *italicized text* and underlined text.

You can also use HTML tags to display text using ***combinations*** of these styles.

gate through folders on your desktop, you can navigate through the directories and subdirectories on a Web directory.

Yahoo, perhaps the best known and most complete Web directory, provides access to more than 14 main directories and more than 50 subdirectories on its home page (Figure 4.15). If you click on the *Arts and Humanities* directory, you move to a new Web page containing additional directories related to the arts and humanities. You can click, in turn, on any of these directories to further narrow your field of interest on an arts and humanities topic.

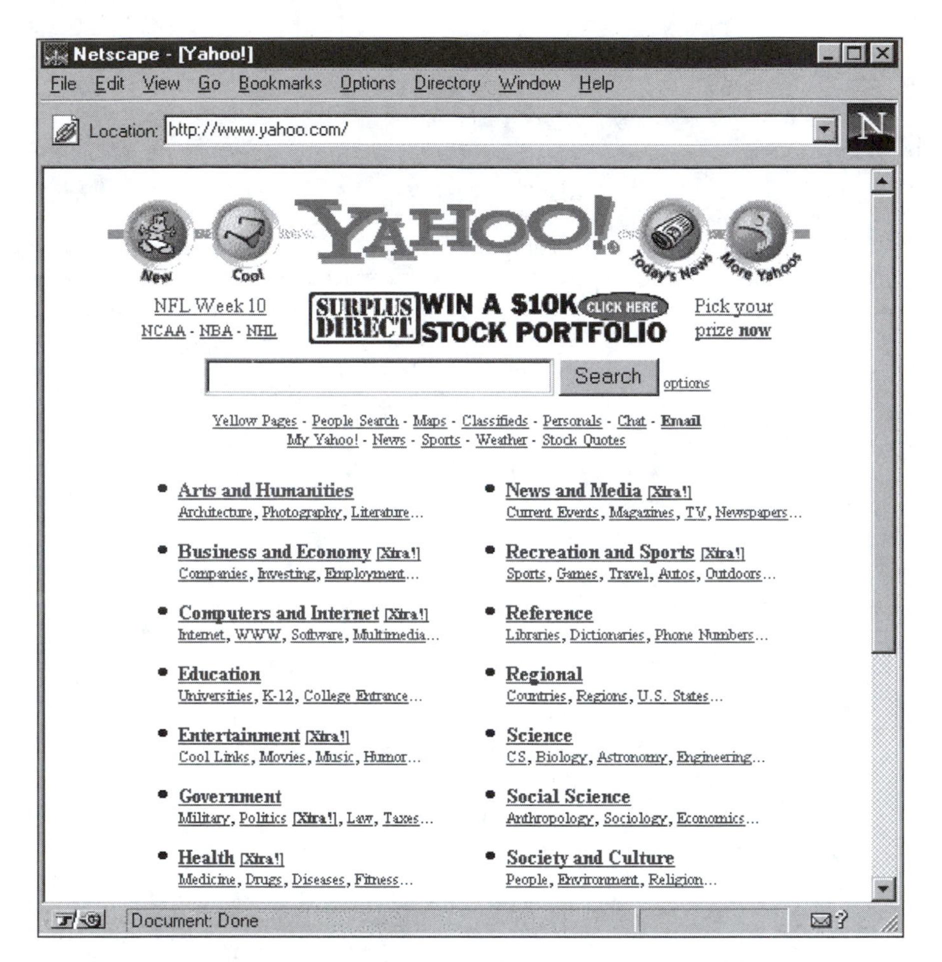

FIGURE 4.15 Directories on the Yahoo home page

Leading Web Directories include:

- The Argus Clearinghouse <http://www.clearinghouse.net>
- Galaxy <http://www.einet.net>
- Magellan <http://www.mckinley.com>
- WebCrawler <http://www.webcrawler.com>
- Yahoo <http://www.yahoo.com>

tip 4.3 ■ Make Note of "Private" Indexers

Some of the best sources of information on the Web are sites containing lists of information on particular topics. Many of these sites are compiled by individuals with strong interests in a particular area—scuba diving or photography, for example. These sites contain links to a range of online materials (most often other Web sites). Other sites are compiled by organizations with an interest in a particular area. For instance, the Alliance for Computers and Writing <http://english.ttu.edu/acw> sponsors a comprehensive list of links on subjects related to the use of computers in the teaching of writing.

You'll often come across private indexes during a search on a particular topic. Bookmark the site or add it to your list of favorites for later reference. These sites are useful for conducting structured searches of the Web (see below).

Searching Combined Search Sites on the Web

Combined search sites, sometimes referred to as super sites, allow you to carry out searches of several Web indexes and directories at the same time. Combined search sites share the same basic principle—a single search can provide you with information from multiple sites. However, they put this principle into practice in distinctively different ways. Leading combined search sites include:

- All4One <http://www.all4one.com>
- Cyber 411 <http://www.cyber411.com>
- Dogpile <http://www.dogpile.com>
- MetaCrawler <http://www.metacrawler.com>
- MetaFind <http://www.metafind.com>
- MetaSearch at Highway 61 <http://www.highway61.com>
- SavvySearch <http://guaraldi.cs.colostate.edu:2000>
- Verio MetaSearch <http://search.verio.net>

All4One uses frames to show the results of searches on AltaVista, Excite, Lycos, and Webcrawler. The frames are resizable, so you can change the relative size of each set of results as you view it.

SavvySearch and MetaCrawler are typical of sites that support search on multiple sites without actually visiting those sites. SavvySearch, originally a master's project for a student at Colorado State University, searches any or all of 25 Web indexes and directories at once. It allows you to specify the number of hits you want from each site. It also allows you to specify whether the results should be integrated into a single list or displayed by search engine. MetaCrawler operates under the same principle, displaying an integrated list of up to 10 results from six leading search sites: AltaVista, Excite, Infoseek, Lycos, Webcrawler, and Yahoo!

tip 4.4 ■ Use Onsite Tools

Search sites such as Lycos and Yahoo can provide you with a great deal of information about relevant Web sites. But they tend to focus on Web sites as a whole, rather than on each page within a site. When you visit a large Web site, such as Microsoft Online <http://www.microsoft.com>, ZD Net <http://www.zdnet.com>, or some government agencies, consider using search tools that have been provided by the designers of the site. This will help you conduct a more focused search of a site that appears to be a promising source of information.

Some of the most common on-site search tools include

- Local search, allowing you to run searches on the site
- Tables of contents, providing an overview of the site
- Site maps, graphical (and sometimes textual) depictions of the contents of the site and its organization
- FAQs, lists of frequently asked questions about the site

ZD Net provides an excellent example of a well designed site that supports local search. From their pages, you can access a search tool, a site map, a list of new information on the site, and a VRML site map (based on the Virtual Reality Modeling Language). In addition, ZD Net provides a table of contents bar along the left side of the screen, further easing your navigation through the site.

Anchored Browsing—Browsing the Web Systematically

It's no accident that the software programs you use to view documents on the Web are called browsers. Browsing—moving from one document to another in a library or bookstore, or for that matter in a shop—is the foundation of activity on the World Wide Web. The concept of linking, which allows readers to move quickly and easily from one document to another, makes it possible for readers to easily browse the Web, often leading them to documents they hadn't known anything about before they started their browser.

In this sense, browsing seems like a more or less random activity. Experienced researchers, however, know that they can browse the Web in ways that increase the likelihood of finding relevant documents. Anchored browsing, sometimes referred to as systematic or branching browsing, builds on techniques used by researchers with print documents. Using works cited lists or bibliographies in print documents that they know are relevant to their search, researchers of print documents use them to locate additional documents. Similarly, you can start with a Web site that you know is relevant to your subject and follow the links from that site to others, branching out as needed to locate information. If a branch plays itself out or moves into irrelevant territory, you can return to an earlier point in the search or to your starting point—the anchor of your search.

In addition to using specific Web documents as a starting point, you can also use this technique in conjunction with searches of Web indexes and directories.

> **tip 4.5 ■ Don't Ignore Your Browser Software**
>
> Your Web browser provides a number of tools that can help you locate and organize information. Like word processing programs, browsers allow you to locate information on a page using the **FIND** command. You can also easily organize information as you search the Web by creating, organizing, and annotating bookmarks (in Netscape Navigator) and favorites (in Microsoft Internet Explorer). You'll find the ability to organize bookmarks and favorites in folders particularly useful. To learn how to work with bookmarks and favorites, consult your browser's online help.

Searching GopherSpace

Since you can use leading browsers to navigate through Gopher sites, you might expect that you could locate information in GopherSpace using Web indexes and directories. But this is not necessarily the case. A great deal of useful information continues to be available via Gopher, and nowhere else.

The best place to locate information in GopherSpace is via Veronica <gopher://veronica.scs.unr.edu:70/11/veronica>, available from most Gopher sites (Figure 4.16). Veronica allows you to conduct keyword searches of titles of folders and files in GopherSpace. It does not support a full text search of the files themselves, but it does support Boolean searches, wildcards, and limits on types of material. To learn more about searching on Veronica, consult the document "How to Conduct Veronica Queries," available at <gopher://gopher.scs.unr.edu:70/00/veronica/how-to-query-veronica> or via the Veronica gopher menu.

Most Gopher servers allow you to search for information at that institution or to connect to Gopher servers at other institutions. If you know that an institution is a leader in a particular area, check to see if the institution has a Gopher server. If so, search on its server and see what you find.

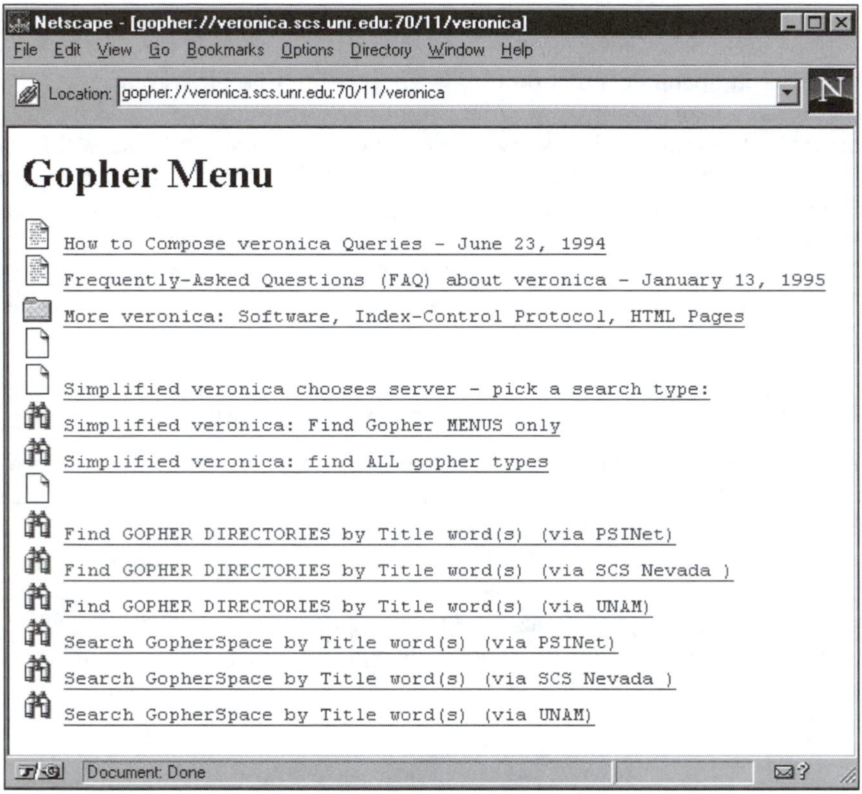

FIGURE 4.16 Veronica, the leading Gopher search site

Consulting Newsgroups and Mailing Lists

Sometimes, the best way to get help is to ask for it. If you are having difficulty locating information on a particular topic, consider subscribing to a newsgroup or joining a mailing list related to that topic.

Before asking for help by posting a message, however, check to see if the group or mailing list has a FAQ—a list of Frequently Asked Questions. If necessary, post a message to the group or list to find out if the FAQ exists and where it can be located. People usually don't mind responding to questions about whether a FAQ exists, but posts of questions addressed in the FAQ, if one exists, are considered poor etiquette—wasting people's time with questions for which answers are already available.

If your question isn't addressed in the FAQ, consult the archives of the group and, if it's not answered there, consider posting your question. Because you're dealing with people rather than with texts, you might find it advantageous to spend some time reading previous posts to get a sense of how people ask and respond to questions on the group or list. If you don't need an immediate response, you can also participate in the group or mailing list without posting a message. This practice, called lurking, is a good way to get a sense of the dynamics of the group or mailing list.

Lurking can also help you identify members of the list or group who seem most knowledgeable on your subject. After lurking for a while, you may find it most expedient to send an electronic mail message directly to a member of the group or list, rather than post to the group or list.

If you ultimately post a question to the group, consider introducing yourself in a sentence or two (your reading of the posts sent to the group will let you know whether this is customary in the group). Make sure that you include your name in your post and your electronic mail address. Some members of the group or list may want to reply to you directly, rather than to the group or list.

EVALUATING ONLINE INFORMATION

As we noted at the beginning of this chapter, locating information is becoming less of a challenge than it once was. The real challenge, it seems, is evaluating the information you find. Unlike much of the information found in a library, online information is not screened first for worth. It can be difficult to determine, for instance, whether a post to a newsgroup was made by a respected expert in the field or someone who wouldn't recognize a fact if it fell in his lap. It can be equally difficult to determine if the information found on a Web site is objective or put there to serve a particular purpose, such as advocating a cause or persuading you to purchase a product.

As the amount of information available via the Internet, commercial online services, and local area networks has grown, writers, librarians, and teachers have struggled to develop useful evaluation criteria. The most commonly used criteria focus on document's author, publisher, purpose, source of information, accuracy and completeness of information, and the timeliness of the document and its information.

The Author of a Document

Despite claims by some literary theorists, authorship matters—at least when it comes to using documents as sources of information. One of the first questions you'll want to ask about a document is, "Who wrote it?" Knowing an author's name, credentials, and organizational affiliation (if any) can help you learn whether an author:

- Has formal training or experience in a particular area
- Has published on this subject elsewhere
- Has had his or her work reviewed by critics or members of the field
- Is frequently referred to or linked to by other authors

If contact information is included with the document, you can request information about an author's qualifications by calling or sending electronic mail to the author.

Unfortunately, it can be difficult to determine who wrote some documents on the Internet. Web pages, for instance, can be published by anyone with access to an institutional or company Web site or by anyone with enough money to pay for Internet access. Sometimes, you can't even trust a post to a newsgroup or mailing list. The author may be using an alias or another person's account.

In cases where you can't determine the authorship of a document, consider requesting information from the organization (if any) that sponsors a Web site, Gopher site, newsgroup, or mailing list.

The Publisher of a Document

In many cases, the publisher of an online document matters at least as much as the document's author. If you find information on a government Web site, for instance, you might be inclined to trust it more than information on the home page of someone you've never heard of. Similarly, your position on a particular issue can shape your perceptions of the trustworthiness of information found on Web sites, Gopher sites, newsgroups, or mailing lists affiliated with particular organizations. If you favor gun control, for instance, you might not trust every document you read on the National Rifle Association's home page <http://www.nra.org>.

Although newsgroups and mailing lists typically don't have publishers, they tend to represent a particular set of interests. If you're a fan of the Macintosh computer, for instance, you might not put a great deal of trust in a post entitled "The Best Mac is No Mac" on the newsgroup comp.os.ms-windows.advocacy.

To learn more about a particular publisher:

- Look for organizational information listed on the document
- Look for the affiliations of a document's author (if any are listed)
- Send electronic mail to a Web site's Webmaster (if one is listed)
- Consult a newsgroup or mailing list FAQ (frequently asked questions)
- Search the Internet for information about a publisher

The Purpose of a Document

Always ask why a particular document has been written. As an example, if you're thinking about purchasing a new computer, you'll probably be able to locate a great deal of information online. Some of that information might be found in documents on a Web site sponsored by the company who makes the computer. More information might be found in documents at Web sites sponsored by various computer magazines—sometimes in paid advertisements and sometimes in reviews. Still more information might be found on newsgroups that deal with the particular kind of computer.

The purpose of each of these documents—to persuade you that a particular computer is right for your needs, to inform you about the computer's performance and features, and to share experiences and opinions about the computer—should be an important part of your evaluation. You'll find that the documents you read at Consumer Reports Online <http://www.consumerreports.org>, for example, have a much different purpose than those you read on a Web site sponsored by a company that exists to sell computers.

It's not always easy to determine the purpose of an online document. Sometimes, documents that appear informative are actually persuasive. To determine a document's purpose:

- Read the document for evidence of authorial or publisher bias
- Consider the nature of the organization publishing the document
- Consider the amount and kind of information used in the document

The Source of Information in a Document

Few experienced researchers take at face value the information they find in a print document. One of the first things they do is to attempt to determine where the information came from. If it's based on primary research, as is the case in many articles in scholarly journals, they will assess the methods used to conduct the research. If it's based on the research of others, they'll look for a works cited page, in-text citations, or a bibliography to see where the information came from. If they're thorough, they'll check those sources as well.

Researchers should apply the same techniques to online documents:

- Ask whether a document attributes its sources of information, either through in-text citations, links to other documents, or works cited pages.
- If a document contains references to other documents, check them out. If the document treats some information as common knowledge in a field, consult other sources in the same field.

- Does the author of a document appear to be aware of key questions and controversies in a field? Are these questions or controversies acknowledged or addressed?
- Think carefully about the methods used to collect information included in a document. Does the author report on how the information was collected? If so, does the author follow established methods of information collection? Does the author appear to have been fair and unbiased in reporting and interpreting the results?

The Accuracy and Completeness of Information in a Document

An old joke recounts the experience of a rookie sales representative who leaves a sales meeting in awe of his more experienced colleague. "Where did you find those figures?" the rookie asks, clearly impressed by his colleague's ability to make the sale. "I made them up," said the other sales representative.

It's easy to be taken in by inaccurate information—or even by outright fabrications—in documents of any kind. Online documents, which often haven't been reviewed before publication, can contain inaccuracies, omissions, and outright lies. As a careful researcher, you should try to:

- Locate any primary sources of information used in a document
- Determine whether other documents on the subject make use of the same information (this builds on the principle of safety in numbers, but it also lets you know whether the information is accepted by other authors working with the subject)
- Determine whether the document includes a reasonable amount of supporting information
- Determine whether the document contains any inconsistencies that suggest an attempt to mislead or bias the reader

The Timeliness of Information in a Document

Sometimes, old information is worse than no information. If you're working on an issue in which information changes on a weekly or even daily basis, such as voters' perceptions of a political candidate, make sure that your information is as up to date as possible. A poll conducted two months ago is of little value if you want to report current thinking about an issue or candidate.

If you are working on an issue in which timeliness is important, consider the following techniques when working with online documents:

- Look for creation dates on a document. They are easily found in posts to newsgroups and mailing lists, but may not be included in documents on the Web or on Gopher.

Panel 4.5
Additional Sources of Information about Evaluating Online Documents

As the Web continues to grow, more and more sources will emerge. The sources listed below provide guidance on evaluating Web sites and the information they provide.

ADDITIONAL WEB SITES ABOUT EVALUATING ONLINE DOCUMENTS

Site Name	Author and Affiliation	URL
How to Evaluate a Web Page	Naomi Lederer, Colorado State University Libraries	<http://manta.library.colostate.edu/howto/evalweb.html>
Evaluating Web Resources	Janet Alexander & Marsha Tate, Widener University Wolfgram Memorial Library	<http://www.science.widener.edu/~withers/webeval.htm>
Thinking Critically about World Wide Web Resources	Esther Grassian, UCLA College Library	<http://www.library.ucla.edu/libraries/college/instruct/critical.htm>
Evaluating World Wide Web Information	Ann Scholz, Purdue University Libraries	<http://thorplus.lib.purdue.edu/library_info/instruction/gs175/3gs175/evaluation.html>
Evaluating Internet Research Sources	Robert Harris, Southern California College	<http://www.sccu.edu/faculty/R_Harris/evalu8it.htm>
Evaluating the Quality of Internet Information Resources	Gene L. Wilkinson, Kevin M. Oliver, and Lisa T. Bennett, University of Georgia	<http://itech1.coe.uga.edu/Faculty/gwilkinson/webeval.html>
World Wide Web Site Evaluation for Information Professionals	Susana Alves, Royal Melbourne Institute of Technology	<http://www.bf.rmit.edu.au/Dimals/rguides/website.htm>
Evaluating Information Found on the Internet	Elizabeth E. Kirk, Johns Hopkins University	<http://milton.mse.jhu.edu/research/education/net.html>
Evaluating Information on the Internet	Shawn Patterson, Alan Wendt, and Robert Schroeder	<http://www.udmercy.edu/htmls/Academics/library/evaluati>
Ten C's for Evaluating Internet Resources	Betsy Richmond, University of Wisconsin-Eau Claire, McIntyre Library	<http://www.uwec.edu/Admin/Library/10cs.html>

- Look for a copyright notice. Copyright notices can provide clues about the date of publication.
- Read the document with attention to when information used in the document was collected. If you are working with a research report, for instance, read the methods section for information about when the study was completed.

If you're working with polling data, look for the dates on which the poll was conducted.

- View the directory in which a Web or Gopher document resides. This will provide you with information about when the file containing the document was created or last updated. This is not a foolproof method; recent changes to a document might reflect the date on which a minor revision was made rather than the date on which the document was created. On the other hand, this approach can let you know whether it's been a while since the document was last changed.

LOOKING AHEAD

Operating systems are increasingly blurring the lines between information stored on your computer, local area networks, and the Internet. More important, the amount and quality of information available on the Internet—and in particular on the World Wide Web—continue to grow. Your ability to locate, evaluate, and organize online information, as a result, is likely to become increasingly critical to the success of your writing projects. As you look ahead to your next projects, consider the strategies we discuss in this chapter for working with online information. Like most writers, you'll find some forms of online information easier to work with than others and some strategies for working with that information easier to integrate into your writing process. Given the rapid pace of change on the Internet, however, we strongly suggest you keep up with the way online information is being stored and disseminated.

chapter 5

Finding Writing Support Online

As you become more familiar with the computer, you can turn to a variety of online resources to support your writing ranging from commercial programs to helpful sites on the World Wide Web to communities of writers who offer advice and encouragement to other writers.

In this chapter, we'll discuss online resources that you can use to support your writing. We'll talk about commercial software programs, such as reference materials for writing, utility programs that extend the capabilities of your word processing software, templates that you can use for a range of writing tasks, and tutoring programs. We'll also explore materials available on the World Wide Web, including online textbooks, online writing centers, and dictionaries and style guides, to name only a few. Finally, we'll focus on online writing communities where writers and writing teachers offer advice on a range of issues related to writing.

USING COMMERCIAL SOFTWARE

You can obtain information about commercial software from a wide range of sources, including local software retailers, mail order software retailers, and sites on the World Wide Web. Most commercial writing software falls into one of four categories:

- Reference materials such as dictionaries, thesauruses, encyclopedias, and grammar or style guides
- Utility programs that ease the process of using citations in documents, support the indexing of documents, or provide additional tools for checking spelling, grammar, and style

- Templates for documents such as letters, reports, and articles
- Tutorials that guide you through the process of writing documents

These software programs function either as separate programs or as "add-ins" for your word processing program or for other programs. Separate programs work side-by-side with your word processing program. Add-in programs often appear as a new menu item in a word processing program. They become part of your word processing program.

Reference Programs

Reference programs range from dictionaries and thesauruses to encyclopedias. Usually provided on CD-ROM, these programs serve the same function as a set of reference materials on a nearby bookshelf. Some programs are available separately, including the *Random House Unabridged Dictionary*, *Grolier's Encyclopedia*, and *Microsoft Encarta* (see Figure 5.1).

A number of online reference programs are also published as a set of reference works. *Microsoft Bookshelf*, for instance, includes the *American Heritage Dictionary*, *Roget's Thesaurus*, the *World Almanac and Book of Facts*, the *Concise*

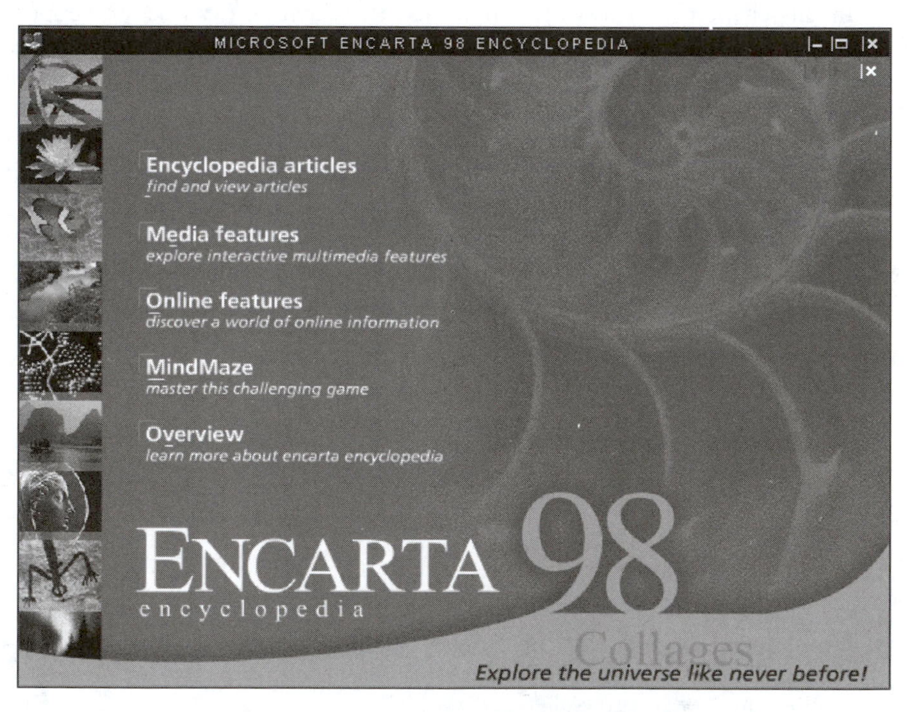

FIGURE 5.1 Microsoft Encarta 98 Encyclopedia Home Page

Columbia Encyclopedia, the *Columbia Dictionary of Quotations*, and other resources. Similarly, *Compton's Reference Collection* includes *Webster's New World Dictionary and Thesaurus*, a book of quotations, an encyclopedia, and five business references including *The Elements of Style*, *The Elements of Business Writing*, a guide to business correspondence, and a resume counselor.

Utility Programs

Utility programs provide features not available in a word processing package. When word processing programs were first introduced, the most commonly used utility program was a spelling checker. When spelling checkers became commonplace parts of word processing programs, the most commonly used utility program was a style and grammar checker. As word processing programs have become more complicated, the need for utility programs has declined.

The most commonly used utility programs include citation databases, indexing tools, and file-format converters. Citation database programs allow you to create and maintain a database of bibliographic citations. Most allow you to assign key words to and annotate your citations. Perhaps most important for writers who find that they must conform to a variety of citation styles (such as the *Chicago Manual of Style*, the *Publication Manual of the American Psychological Association*, or *Modern Language Association Style Manual*, among literally thousands of others), these programs allow you to automatically generate bibliographies in the style you specify. Some of the most popular citation database programs include:

- Niles and Associate's EndNote Plus
- Institute of Scientific Information's Reference Manager
- Nota Bene's Citation
- Personal Bibliographic Software's Pro-Cite
- Research Software Design's Papyrus

As you gain more skill in writing and you produce longer documents, you may find indexing programs, such as Iconovex Corporation's Indexicon, helpful in generating indexes for your documents. Although leading programs such as Microsoft Word for Office 97 include indexing functions, programs such as Indexicon offer a wider range of features.

File-format converters translate files saved in one word processing format to another format. Although your word processing program can convert many formats on its own, file-format converters such as Advanced Computer Innovations' WordPort for Windows and Mastersoft's Word for Word can convert a far wider range of formats.

Some converters are also available to convert from one operating system to another. Acute Systems' TransMac, for instance, allows PCs to read Macintosh diskettes. DataViz' MacLink Plus translates the formats of numerous files between the two platforms.

Another class of add-in utilities consists of programs that extend common word processing features to other programs. hSoft Inc.'s WordScribe, for example, provides an interactive spelling checker and thesaurus for Windows 3.x and Windows 95 applications. It adds a new menu item to these programs and supports on-the-fly correction, customized dictionaries, and common editing commands (e.g., converting tabs to spaces, capitalization, etc.).

Templates

Templates are among the most common types of software used by writers. As the name implies, templates are preformatted, and in some cases prewritten, documents. Templates that focus on formatting issues are similar to those that are provided with many leading word processing packages. HillySun Enterprises' FormatEase and Reference Point Software's Reference Point Templates for APA Style, for example, are templates that format a document in APA style and provide support for entering references in APA style. These programs add a menu category to the word processing program and provide predefined styles in documents based on the templates.

"Fill-in-the-blank" templates support both formatting and entering information into a document. WritePlace Software's Writing Coach for the Mac, for example, uses what its creators call "think-in-the-blank" worksheets to "break the writing process into 'bite-sized' pieces." Similarly, Elfin Forest Software Group's Computerized Thesis Writing Guide helps thesis writers construct a thesis by filling in the blanks in a thesis template.

Some templates go beyond formatting and "fill-in-the-blank" support, providing formatted documents that you can use or modify as needed. WriteExpress Corporation's WriteExpress Interactive Business Letters, for instance, provides 1,650 business and personal letters as well as step-by-step instructions on over 500 writing topics. Similarly, Avantos Performance Systems' Review Writer 3.0 allows you to rate an employee's performance in specific areas and then suggests prewritten text that describes the rating you've provided.

Tutorials

Tutorials can help you generate ideas, organize information, and revise text, among other common writing activities. Some tutorial programs, such as Prentice Hall's Writer's Helper, a pioneering invention and revising tool developed by William Wresch, support a wide range of writing activities. Originally developed for students in college writing courses, the latest version of Writer's Helper supports common business writing tasks.

Other tutorial programs focus on one specific aspect of the writing process. IdeaFisher for Windows or Macintosh, for instance, focuses on generating ideas for stories, products, company names, advertising, marketing, and commercial products.

Still other tutorial programs focus on particular kinds of documents. A wide range of tutorial programs are available for playwrights and screenwriters, among them Collaborator for DOS or Macintosh, Screenplay Systems' Dramatica Pro for Windows or Macintosh, and Blockbuster for Windows or Macintosh.

FINDING SUPPORT ON THE WORLD WIDE WEB

The World Wide Web provides a vast array of resources for writers, including dictionaries and glossaries, thesauruses, encyclopedias, grammar guides, style and citation guides, online texts about writing processes and genres, and online writing centers. You can find a comprehensive list of these materials at the Online Writing Center at Colorado State University <http://www.colostate.edu/Depts/WritingCenter>.

Dictionaries and Glossaries

The number of dictionaries and glossaries on the Web is growing rapidly. These sites function like commercial software dictionaries, although they do not yet provide direct integration into word processing programs. Some of the most useful dictionaries and glossaries on the Web include:

- Merriam–Webster Dictionary
  <http://www.m-w.com/dictionary>
- Dictionary.com
  <http://www.dictionary.com/>, which provides links to a range of resources including dictionaries, thesauruses, and glossaries

You can also find specialized dictionaries and glossaries on the Web. A representative sample includes:

- WWLIA Legal Dictionary
  <http://www.wwlia.org/diction.htm>
- NetLingo
  <http://www.netlingo.com>, a dictionary of Internet and Web terms
- Semantic Rhyming Dictionary
  <http://www.cs.cmu.edu/~dougb/rhyme.html>
- NetGlos—The Multilingual Glossary of Internet Terminology
  <http://wwli.com/translation/netglos/netglos.html>

Thesauruses

Thesauruses are also available on the Web. Some of the most useful thesauruses on the Web include:

- Merriam–Webster Thesaurus
  <http://www.m-w.com/thesaurus>
- Roget's Internet Thesaurus
  <http://www.thesaurus.com>

Encyclopedias

Currently among the least common writing resources available on the Web, encyclopedias are likely to become more widely available. A brief list of encyclopedias on the Web includes:

- The Free Internet Encyclopedia
  <http://clever.net/cam/encyclopedia.html>
- The Cyberlaw Encyclopedia
  <http://gahtan.com/techlaw/>
- The Web Developers Virtual Library
  <http://wdvl.com/>
- Microsoft Encarta Online
  <http://encarta.msn.com/EncartaHome.asp>

Grammar Guides

Grammar guides are among the most common writing resources available on the Web. These sites range from online lessons in grammar to information about grammar "hotlines" that allow you to submit a question to a grammar expert. Representative sites include:

- Online English Grammar
  <http://www.edunet.com/english/grammar/index.html>
- Grammar and Style Notes
  <http://www.english.upenn.edu/~jlynch/Grammar>
- Grammar Hotline Directory
  <http://www.tc.cc.va.us/vabeach/writcent/hotline.html>

Style and Citation Guides

Style guides provide information about writing issues such as voice, tone, and paragraph and sentence structures. The most widely accessed Web site dealing with style in general writing is the 1918 version of William Strunk, Jr.'s *Elements of Style* <http://www.columbia.edu/acis/bartleby/strunk>. Although much newer editions of the book are available in print, many writers find the advice available on this site useful. Sites dealing with citation style most often focus on commonly used styles, typically those established by the Modern Language Association and the American Psychological Association. These sites address such issues as the preparation of a works cited page and in-text citation. Often, these sites focus primarily on or include information about citing online texts.

Representative sites dealing with style and citation include:

- Frequently asked questions about the Publication Manual of The American Psychological Association
  <http://www.apa.org/journals/faq.html>
- MLA Style
  <http://www.mla.org/main_stl.htm>
- APA Style Citations of Electronic Sources
  <http://www.cas.usf.edu/english/walker/apa.html>
- MLA-Style Citations of Electronic Sources
  <http://www.cas.usf.edu/english/walker/mla.html>
- Beyond the MLA Handbook: Documenting Electronic Sources on the Internet
  <http://eagle.eku.edu/honors/beyond-mla/>
- Chicago Style Guide
  <http://healthlinks.washington.edu/hsl/infoguid/chicago.html>

Online Resources for Writing Processes and Genres

These resources are less common than those listed above. Often, you will find resources of this type on Online Writing Centers or Online Writing Labs (OWLs) discussed in the next section. These resources range from brief handouts to extensive documents incorporating graphics, audio, and video. You can access some of these resources at:

- Reference Materials on the Online Writing Center at Colorado State University
  <http://www.colostate.edu/Depts/WritingCenter/reference.htm>
- St. Cloud State University's Literacy Education Online
  <http://leo.stcloud.msus.edu>
- Paradigm Online Writing Assistant
  <http://www.spaceland.org/paradigm>

Online Writing Centers and Labs

Online writing centers and labs (often referred to as OWLs) offer online support for writers. Most OWLs are affiliated with colleges and universities, but some commercial sites have recently begun operating. Although the majority of OWLs exist primarily to provide information about campus writing centers, several provide extensive support on writing issues including writing processes, genres, grammar, style, and Internet-based research. The Online Writing Center at Colorado State University <http://www.colostate.edu/Depts/WritingCenter>, for instance, provides several thousand pages of information on topics ranging from considering audience and purpose, to using electronic mail, to writing technical reports in civil engineering (see Figure 5.2).

You'll find links to many other OWLs on the "Other Online Resources" pages of the Online Writing Center at Colorado State University. Some of the most comprehensive sites include:

- Purdue University's Online Writing Lab
  <http://owl.english.purdue.edu>
- The Online Writery at the University of Missouri-Columbia
  <http://www.missouri.edu/~writery>

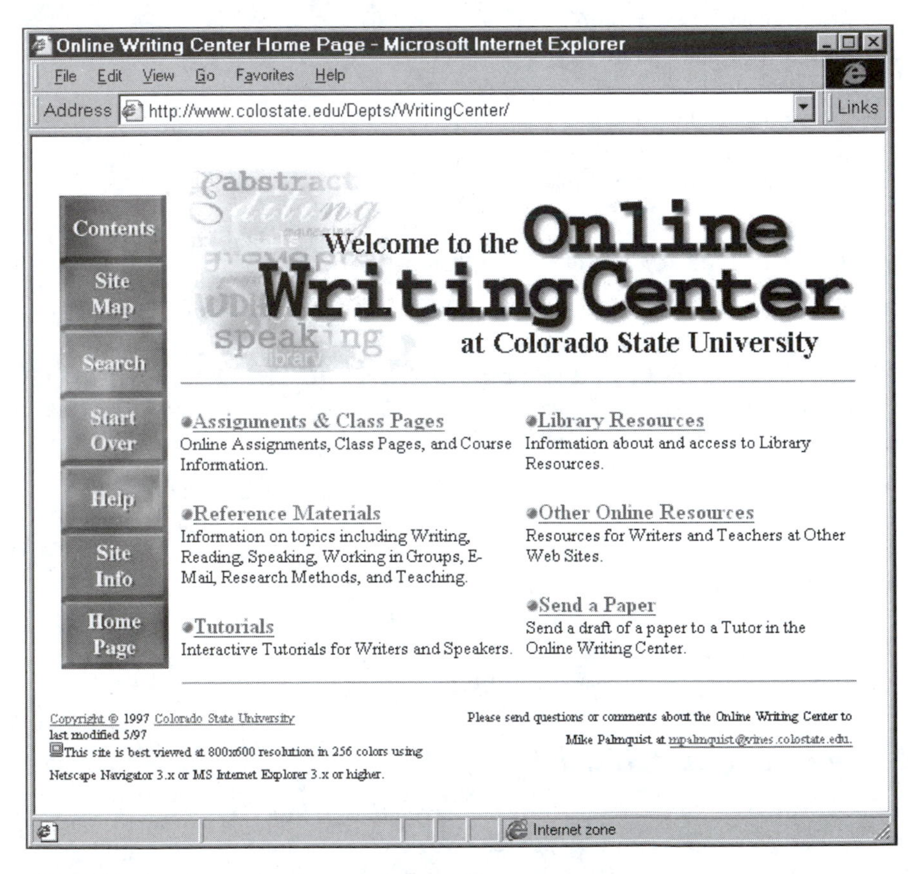

FIGURE 5.2 Example of an online writing center

- The Writers' Workshop at the University of Illinois at Urbana–Champaign <http://www.english.uiuc.edu/cws/wworkshop/writer.html>
- The University of Michigan OWL <http://www.lsa.umich.edu/ecb/OWL/owl.html>

Many OWLs provide access to online feedback and tutoring. Using electronic mail or a form on a Web page, writers can send their drafts to experienced tutors and receive a response via electronic mail. Typically, however, this support is offered only to writers affiliated with the college or university where the OWL is located. Some commercial sites also provide online tutoring and feedback on a fee basis. Before you use such sites, check them out as you would any company providing commercial services. And check with the writing center or composition program at a local college or university and ask for a recommendation of experienced writing teachers, tutors, or editors.

If you choose to seek support from a commercial service, however, you won't find it difficult to locate them. A comprehensive list is available on the Yahoo search site <http://www.yahoo.com> under the category BUSINESS AND ECONOMY:COMPANIES:COMMUNICATIONS AND MEDIA SERVICES:WRITING AND EDITING:EDITING AND PROOFREADING. If the idea of an online writing center or a commercial service doesn't appeal to you, investigate some of the online writing communities discussed in the next section.

JOINING ONLINE WRITING COMMUNITIES

Online writing communities come in a variety of forms and serve a number of purposes. Writers form communities via electronic mail discussion lists, newsgroups, Web sites, Chat channels, and MOOs (we'll talk about each of these below). Some of these communities focus on particular kinds of writing, such as poetry, fiction, drama, technical writing, or business writing. Often, these communities draw members from around the globe. TECHWR-L, for instance, an electronic mail discussion group that explores issues related to technical writing, has several thousand members from both inside and outside North America. Other lists and Web sites help connect writers interested in working on children's literature, romance novels, mystery, and science fiction, among others. A comprehensive list of these communities can be found on the Writer's Resource Page <http://www.arcana.com/shannon/writing.html> or on Yahoo's SOCIAL SCIENCE:COMMUNICATIONS:WRITING listings <http://www.yahoo.com>.

In addition to online communities that form as a result of common interests in a topic, numerous online communities have emerged as extensions of established writing organizations in various cities or regions. The Loft, a Twin Cities-based literary group, has created a Web site that helps keep members in touch with each other and informed about upcoming events <http://www.tc.umn.edu/nlhome/m555/loft/index.html>. Similarly, the Writers Center of Greater

Cleveland <http://www.cleveland.com/ultrafolder/litlife/writing/bio.html> provides a focal point for writing activity in that city, as well as opportunities for writers to interact electronically.

With so many online writing communities, deciding which ones to join can be a daunting task. To give you a sense of these communities, consider the representative lists below which focus on mailing lists and newsgroups for writers, Web sites for writers, and Chat channels and MOOs for writers.

Mailing Lists and Newsgroups for Writers

Mailing lists and newsgroups are types of electronic mail discussion groups (see Chapter 4). The primary differences between them are:

- Mailing lists send messages directly to your electronic mail account while you need to use a newsgroup reader to view messages sent to a newsgroup
- You must join a mailing list (so that the list server knows which electronic mail addresses to send messages to) while you must subscribe to a newsgroup so that you can read it using your newsgroup reader

You'll find tens of thousands of mailing lists and newsgroups on the Internet. Fortunately, locating mailing lists and newsgroups of interest to you isn't difficult. Several Web sites provide information that can help you decide which groups to join:

- Liszt, a searchable mailing list directory, contains information about more than 70,000 electronic mail discussion groups
  <http://www.liszt.com/>
- Deja News, a searchable list of newsgroups and newsgroup posts, allows you to search or browse its listings
  <http://www.dejanews.com>
- Yahoo provides a list of sites you can search for newsgroups and newsgroups postings
  <http://www.yahoo.com/News_and_Media/Usenet/Newsgroup_ Listings>

Well established mailing lists dealing with writing include:

- ACW-L, the mailing list of the Alliance for Computers and Writing. This list focuses on the use of computers to support writing and writing instruction. For more information, visit the ACW-L Web site at
  <http:// english.ttu.edu/acw>.
- TECHWR-L, an open list for discussion of all technical communication issues. You can read the archives of this list at
  <http://listserv.okstate.edu/archives/techwr-l.html>.
- NASW-TALK is the mailing list for the National Association of Science Writers. For more information about this list, send electronic mail to cybrarian@nasw.org.

Representative newsgroups dealing with writing issues include:

- misc.writing
- misc.writing.screenplays
- rec.arts.poems
- rec.arts.poety
- rec.arts.prose
- rec.arts.int-fiction

These newsgroups draw from a large population and differ from the more academic posts typically found on the mailing lists mentioned earlier.

Web Sites for Writers

Although you could search for Web sites that contain the word "writing," the sheer volume of hits will lead you to pursue other strategies (a July 1998 search on HotBot, for instance, yielded 2,037,634 hits). One approach is to explore the categories found on Yahoo <http://www.yahoo.com>. By progressively narrowing your search, you can locate sites that interest you.

In addition to the Web sites for the Loft and the Writers Center of Greater Cleveland, you may find some of these sites of interest:

- Alien-Flower, a site where "poetry-lovers practice poetry" <http://www.sonic.net/web/albany/workshop>
- The International Poetry Forum <http://www.artspa.com/poetry/>
- The Slush Pile, a site for writers of children's books <http://www.theslushpile.com>
- The Society for Technical Communication <http://www.stc-va.org>
- The Mystery Writers' Forum, a threaded discussion forum on the Web <http://www.zott.com/mysforum>
- The Screenwriters and Playwrights Home Page <http://www.teleport.com/~cdeemer/scrwriter.html>

Chat Channels and MOOs for Writers

Chat channels and MOOs (Multi-User Dungeons—Object Oriented) allow people to type messages to each other in real time, even though they might be separated by thousands of miles. The messages appear on the screen in the order they are written.

Chat rooms and MOOs allow writers to share ideas, to collaborate on projects, and to ask for information about particular topics. Many writing classes use Chat rooms and MOOs to carry out collaborative brainstorming sessions, to critique drafts of papers, and to engage in role-playing games, such as Devil's Advocate.

In general, Chat rooms and MOOs provide writers with the opportunity to talk about their writing with other writers.

Increasingly, writers are using Chat and MOOs to meet and discuss writing issues online. To join the discussions in a writing-related Chat room or MOO, check out the following:

- MediaMOO at purple-crayon.media.mit.edu: 8888
- LambdaMoo at lambda.moo.mud.org: 8888
- LinguaMOO at lingua.utdallas.edu: 8888 or visit the LinguaMoo Web site <http://lingua.utdallas.edu/>
- Diversity University at moo.du.org: 8888 or visit the DU Web site <http://www.du.org>

LOOKING AHEAD

The idea of the writer as a solitary individual working for unbroken stretches of time on a document is, with some rare exceptions, inaccurate. Most writers spend as much time talking about their writing projects—generating ideas, collecting information, and soliciting advice, among other things—as they do writing text. Increasingly, writers are also turning to online interaction, on local area networks and on the Internet, for support for their projects. As you look ahead to your next project, consider ways in which you might find online support for your writing. Like a growing number of writers, you may be pleasantly surprised by what—and who—you find waiting for you.

chapter 6

Writing Faster with a Computer

Some things are so enjoyable that we wish they could go on and on. It could be skiing the back country on a cloudless day high in the Rockies. It could be reading a book that transports us to another place and time. It could be a perfect day of fishing or an outing with friends. And, for a select group, it could be sitting down to write a report, an essay, a novel, or a short story.

If you want your writing experience to go on and on, skip this chapter—you won't want to learn how the computer can reduce the time you spend writing, and you won't be interested in strategies that can help you become a more efficient writer.

If you're like most writers, however, read on. In this chapter, we explore how you can reduce the time you spend writing, revising, editing, and formatting your text. We begin with a brief review of composing strategies discussed in earlier chapters. Then we turn to word processing tools and techniques that can dramatically reduce the time you spend on specific writing activities. We conclude the chapter with a discussion of how you can use and reuse (or recycle) text in multiple documents, a technique that can dramatically reduce the time you spend writing.

COMPOSING ON THE COMPUTER

The first chapters of our book focused on how computers, and in particular word processors, can help you compose more efficiently and effectively. In those chapters, we explored how the computer can support a variety of writing processes. In this chapter, we focus primarily on using your computer to save time. Rather than duplicate discussions of specific commands and tools, we sometimes refer to information found in earlier chapters.

Recall that composing begins with generating ideas and collecting information. In Chapter 1, we discussed several techniques that can be useful for generating ideas, among them blind writing, looping, and outlining. We also discussed several techniques for collecting and organizing information, which we built on in Chapter 4. And we discussed, in Panel 1.1, the efficiencies that can be gained by learning to touch type. Additional sections of the book that can help you compose more effectively and efficiently include discussions of drafting, planning, reviewing, and revising in Chapter 1, of editing in Chapter 2, and of designing and formatting documents in Chapter 3.

Consider, for instance, how much more quickly you can do a simple idea-generating activity, such as looping, on the computer than on paper. Looping involves writing quickly about an idea, then using a short passage from the text you've just generated as the starting point for another writing session. Doing something as simple as copying the passage and pasting it into a new document saves time over copying it out by hand onto a new piece of paper. Similarly, creating an outline is much more efficient on a computer, which allows you to easily add, delete, or rearrange text.

Many writers will save a great deal of time using the computer to collect and organize information. Writing information by hand on note cards—a technique discussed in many writing textbooks—is far more tedious than working with it on a computer, where you can store the results of entire searches, create templates to organize information from your research, and easily rearrange and recategorize information as your writing needs change.

Other writers will save time using spelling checkers, grammar and style checkers, and the **FIND** and **REPLACE** commands to help edit their texts. Still others will save time using formatting and design tools. A good way to determine how you might save the most time writing is to reflect on your strengths and weaknesses as a writer and then reconsider our discussions in earlier chapters about composing processes, design processes, and information-gathering activities. You're likely to discover several techniques that will enhance your writing efficiency and effectiveness.

Using Commands and Tools to Save Time

If you are writing a longer document, you can save time through effective use of formatting commands and tools. Similarly, if you write a number of shorter documents that share formatting features, you can save time over several writing projects. The key is understanding the range and application of these commands and tools. In earlier chapters, we introduced and discussed techniques related to formatting text and designing documents. Here, we'll focus on how you can use formatting commands and tools to speed your writing.

Styles and Document Templates. Styles and document templates are particularly useful when you want to quickly define or redefine the design of a particular document. We explored the STYLES command and document templates in detail in Chapter 3 (Panel 3.10). The STYLES command allows you to define and apply several font and paragraph formatting commands at once. Document templates contain a set of styles that can be used to create particular kinds of documents.

If you are creating a new document, you can use a document template to define its look and feel. Many word processors provide a set of document templates. Typical template topics include memos, letters, faxes, resumes, reports, manuals, brochures, purchase orders, theses, and Web pages. You can modify these templates for your own uses or create new document templates. To learn more about using templates with your word processor, consult its manual or online help.

The STYLE command allows you to create new styles, modify existing styles, and copy styles from one document or document template to another. One of the quickest ways to change the look of your document is to redefine a style (Figure 6.1). Changing the definition of a style will change all text to which the style has been applied. As we noted in Chapter 3, this is a useful technique for exploring alternative document designs. It will seem even more useful, however, if you need to change the look and feel of your document, particularly if you're working on a deadline.

Menu Commands, Button Bars, and Keystroke Commands. Many writers who use computers use menus to carry out most commands. To cut or copy text, for example, they will click on the Edit menu and then on the CUT or COPY menu command. This involves six steps: moving your hand from the keyboard to the mouse, moving the mouse pointer to the menu bar, clicking on edit, moving the mouse down the menu, clicking on the command, and then moving your hand back to the keyboard. You can reduce the amount by one step if your hand is already on the mouse, but it's still a lot of movement for a relatively simple command. Multiply those movements—and the few seconds it takes to carry them out—by a factor of 100 or more and you'll find that you're spending a lot of time invoking commands with the mouse.

You can save time with simple editing and formatting commands in three ways: you can use the keyboard to invoke commands; you can use icons on the button bar; and, in most word processors, you can use the right mouse button to invoke context-sensitive menus. You can save the most time by using the keyboard. In most Windows word processing programs, for example, the keyboard command for CUT is Ctrl-x (see Appendix B for a list of common keyboard commands).

Taking the time to memorize keyboard commands for common editing and formatting commands will help you write more quickly. But even the most proficient computer users won't memorize every command. Using icons on the

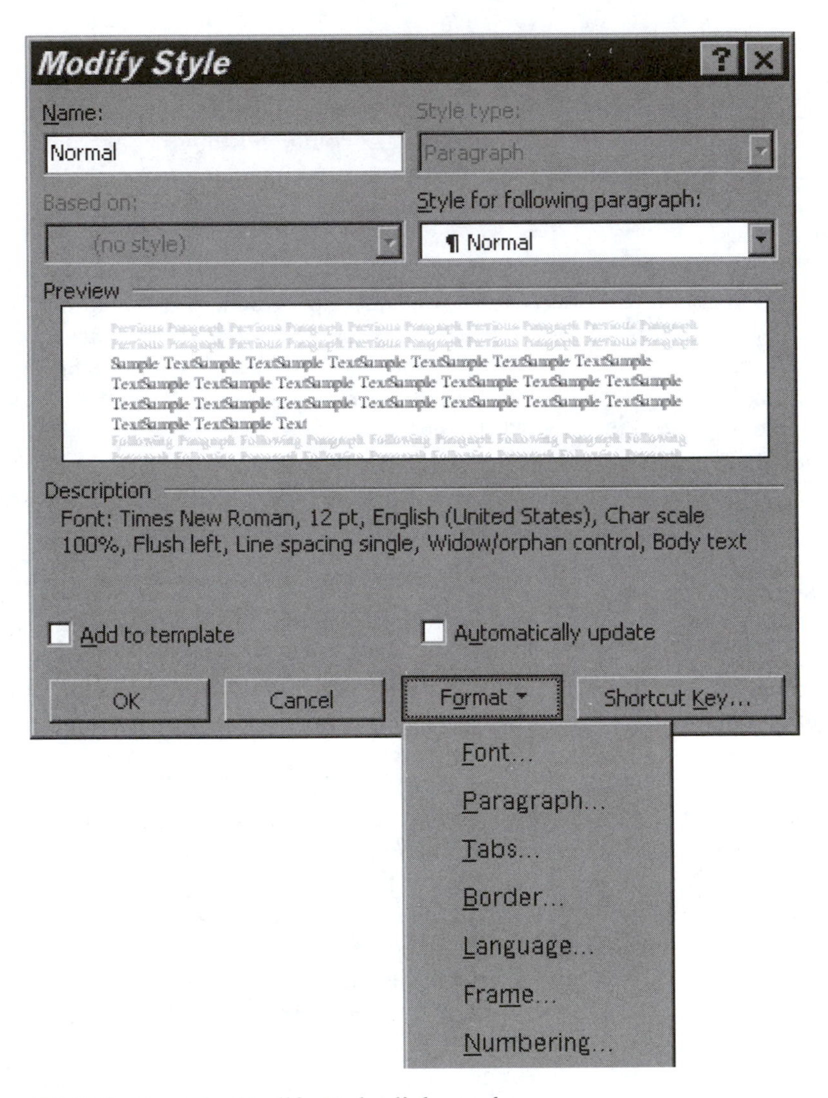

FIGURE 6.1 The Modify Style dialogue box

button bar, although not as efficient for simple commands as using the keyboard, provides an alternative to using menu commands. Icons on button bars provide easy access, both physical and visual, to a range of common commands. If you are unfamiliar with the function of a particular icon, you can move your mouse over the icon and, in most word processing programs, view a flag that will tell you what the icon does. This helps you learn the functions of individual icons and helps you save time when you need to locate an unfamiliar command.

You can also save time by invoking commands through the right mouse menu. In most leading word processing programs, clicking on the right mouse button pops up a context-sensitive menu—that is, a menu containing a limited set of commands commonly applied to the kind of text or graphic that you're working on (Figure 6.2). Similarly, if you're working on a table, the right mouse menu lists commands typically used on text and tables (Figure 6.3). By using right mouse clicks, you no longer need to move the mouse between the text and the menu or the button bar. This reduces your arm or wrist movement, saving additional time and reducing stress on your hands and arms.

For less frequently used commands, such as inserting a table or a bulleted list, consider using the button bar. As we note in Panel 1.3, creating a table using the button bar is the quickest way to create a table in leading word processing programs. Rather than using menu commands to access a TABLE dialogue box, you can use your mouse to create a table containing the appropriate number of rows and columns.

If you find yourself frequently using commands that are not on the button bar in your word processor, consider adding an icon to the button bar or assigning a keystroke command so that you can easily access the command. We discuss customizing your word processor later in this chapter.

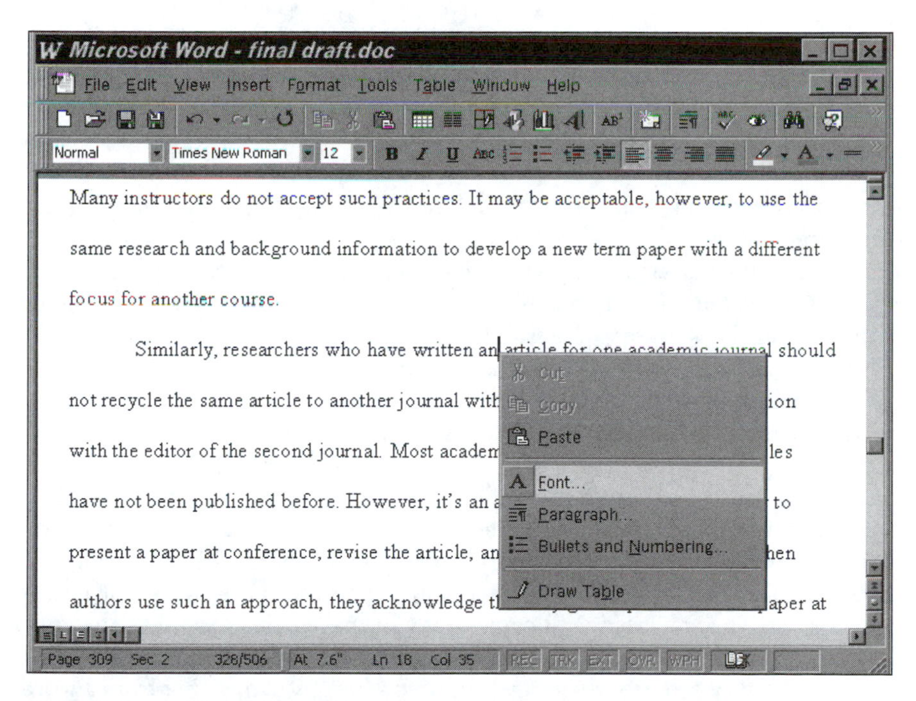

FIGURE 6.2 Right Mouse Menu over text

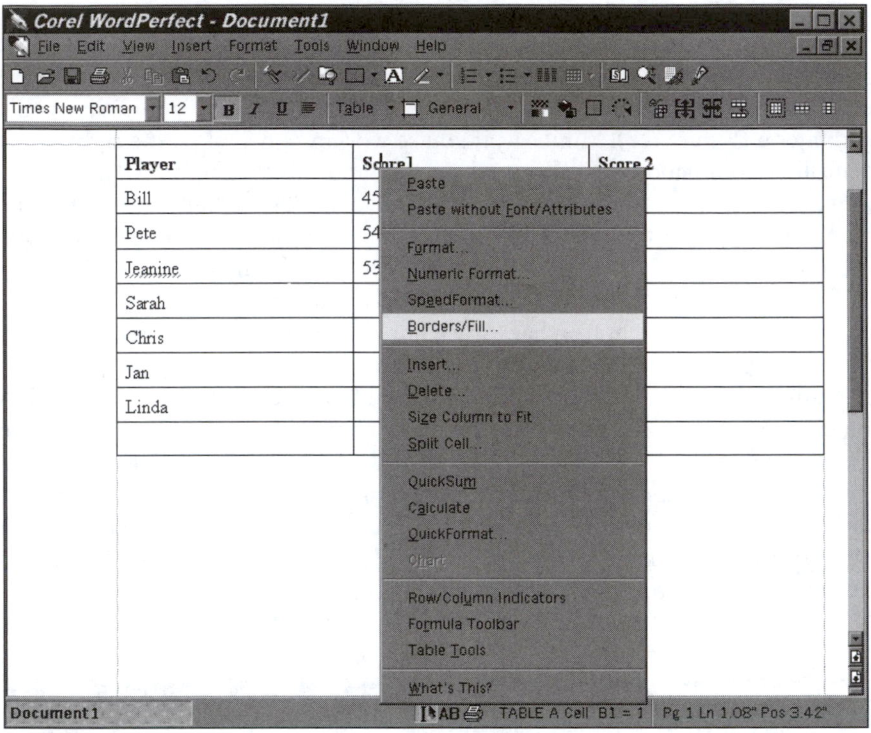

FIGURE 6.3 Right Mouse Menu over a table

Automatic Formatting and Text Tools. You can save time by using automatic formatting and text tools in leading word processors. These tools allow you to concentrate on your writing, rather than on the way your writing appears. They can create bulleted and numbered lists, format headings and paragraphs automatically, change straight quotes into curled quotes, format electronic mail addresses and World Wide Web addresses as hypertext links, and even insert words and phrases when you type acronyms or abbreviations.

Automatic formatting tools allow the word processor to format a document automatically. You can use these tools—referred to using names such as Autoformat, Quickformat, and Check Format—to format a document as you type it or to format a document that you have already created. If autoformatting is turned on, the program will note when you appear to be typing a list and automatically format the text as a numbered or bulleted list. If you use the Autoformat tool to apply formatting to an existing document, it will analyze the document and apply formatting according to a set of rules specified by the designers of the word processor. Some word processors also allow you to copy formatting from a word, selection, or paragraph and apply it to another automatically.

Panel 6.1
Using the Keyboard

Some writers prefer using keystrokes to using the mouse. The CUT, COPY and PASTE keystrokes remain the same across different word processors:

Function	*Keystrokes*
CUT	Ctrl + x
COPY	Ctrl + c
PASTE	Ctrl + v

To use the commands, highlight the copy, then hold down the CTRL key and type the respective letter.

To cut and move a paragraph of text,

1. Highlight the text.
2. Press Ctrl and then the letter x.
3. Move the cursor to where you want to insert the text.
4. Press Ctrl and then the letter v to insert the text.

The word processor should then insert the text in the new location. For other commands with your word processor, look carefully at the edit menu; you'll see a keystrokes list to the right of the EDIT commands (Figure A).

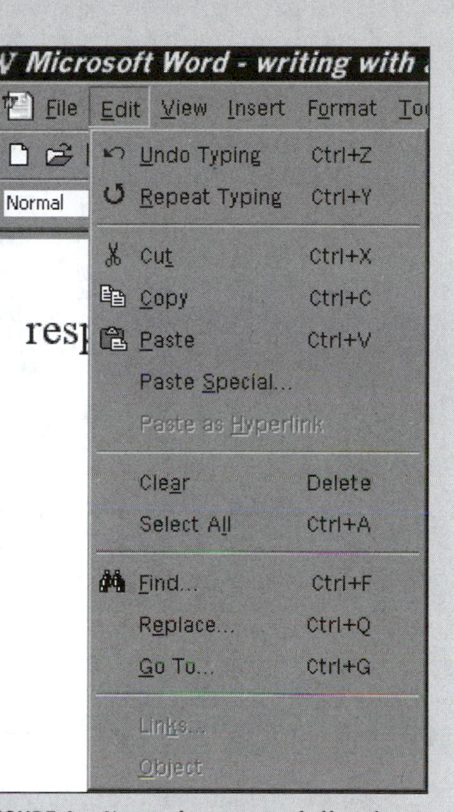

FIGURE A Keystroke commands listed on the Edit menu

Automatic text editing tools replace specific words or symbols as you type. For instance, these tools can automatically capitalize the first words of sentences if you have neglected to do so. They can also substitute particular words or symbols as you type them. For instance, to save time, you might specify that the word *info* should automatically be replaced with *information* each time it is typed. Using this feature, you can specify a wide range of abbreviations and acronyms that can be automatically replaced with specific words or phrases. You can also use the tool to automatically replace commonly misspelled words.

Additional automatic text and formatting options are available in various word processing programs. Among the more common options are correcting two initial capital letters in a word and replacing straight with curled quotation marks. Typically, you can control these options via a dialogue box. To learn more about automatic text formatting and editing tools, consult the online help in your word processing programs (see Figure 6.4).

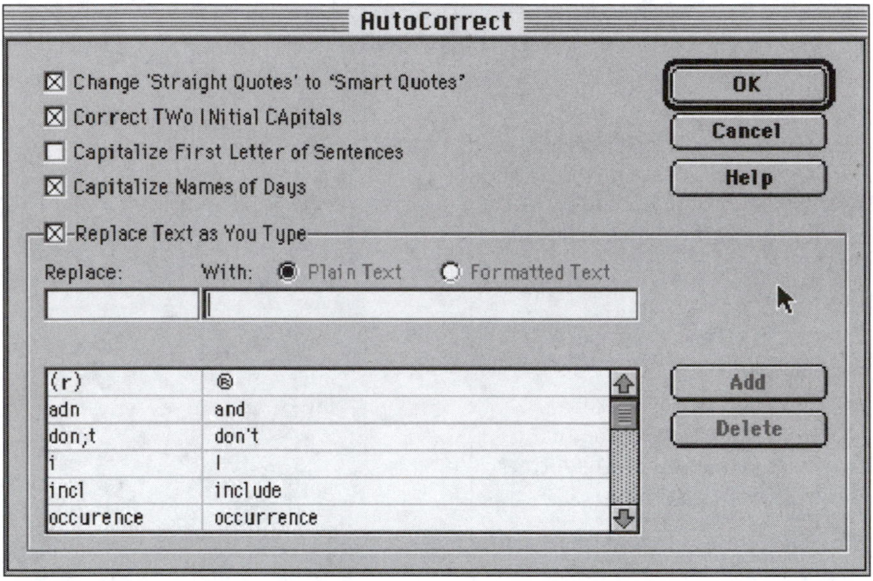

FIGURE 6.4 The AutoCorrect dialogue box

tip 6.1 ■ Speed Typing

To speed your writing, avoid typing long words or phrases—you can use abbreviations and later replace them. Assume you're writing a report on increasing rate of sexually transmitted diseases among adolescents. Rather than typing "sexually transmitted diseases" time and again in your report, you can type *STD* and use the **FIND AND REPLACE** command in your word processor to convert *STD* to *sexually transmitted diseases.* To use this technique:

1. Make a backup copy of your document
2. Move the cursor to the top of the original document

3. Click on **EDIT**
4. Click on **REPLACE**
5. Enter "STD" in the Find What box
6. Enter "sexually transmitted diseases" in the Replace With box
7. Click on the Match Case box
8. Click on the Find Whole Words Only box
9. Click on the **REPLACE ALL** button

The program will replace all occurrences of *STD* with *sexually transmitted diseases* throughout the entire document. When used properly, this process can enable you to speed your writing. For some documents, you can generate several different abbreviations for different words or longer phrases and then use the **FIND AND REPLACE** command. If, however, you are unsure of whether or not you want to replace text throughout the document, use **FIND NEXT** rather than **REPLACE ALL.** Your word processing program will stop at each match. You can check the match and make the replacement if you think it appropriate.

Although this technique works well, keep at least one backup copy of your original files before invoking the global replacement commands. Be sure to use appropriate matching commands so that you replace words or letter strings with exact matches. Using **FIND AND REPLACE** without the proper precautions can cause problems throughout a document. It will replace individual letters within words and make unwanted changes.

If things don't turn out as you'd planned, you can use the **UNDO** command or the **FIND AND REPLACE** command to reverse the changes—for example, you could do a find for the word string and then replace all cases with the abbreviation. Or you can open your backup file, rename it, and try the process again.

Index and Table of Contents Tools. In longer documents, the ability to create and easily update indexes and tables of contents can save you a great deal of time. Simply creating an index is a time consuming chore; finding out that late revisions to a document require that you search for each indexed term and update the page numbers in the index can be extremely dispiriting. Similar challenges face writers who must create and update tables of contents in long documents. If you're working on an undergraduate or graduate thesis, for example, or a long report for a class, you'll find that creating indexes and tables of contents by hand can be quite challenging.

Leading word processing programs, fortunately, offer powerful tools for creating and updating tables of contents. If you have used **STYLES** throughout your document, you can easily generate a table of contents. To learn more about creating and updating tables of contents with your program, consult online help (see Figure 6.5).

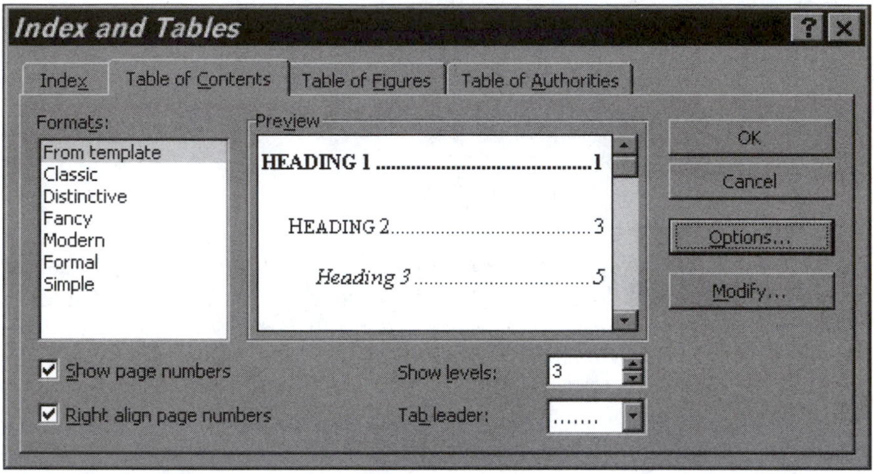

FIGURE 6.5 The Table of Contents dialogue box

tip 6.2 ■ Back Up Document Files

Being paranoid about losing your documents can save a great deal of time. Should you lose a file, retrieving a backup file—a copy of a file that is stored in a different place—and renaming it takes far less time than trying to recreate a document from scratch. Unfortunately, many writers neglect to back up their files because it can be a tedious job and it's easy to forget to do. The costs of losing an important file, however, can be quite high. Rewriting a document because you didn't keep a backup copy wastes your time and keeps you from doing other, more productive things.

At its simplest, backing up a file involves using the **SAVE AS** command in your word processing program to make a second copy of a file on a diskette, network drive, or removable drive. You can also back up files by using utilities that are part of your operating system or by using commercial backup programs. Backup utilities and programs can often be set up so that they run automatically every night or each weekend, when you aren't using your computer.

Regardless of how you back up your files, make sure that you keep your backup files in a safe place—ideally in another room or building. Writers tend to worry about what might happen if the computer breaks down and the files on the hard drive are lost. But files can also be lost as a result of fire or flood. In these kinds of situations, your backups will do you little good if they are stored in the same room as your computer.

Working with Text

When it comes down to basics, word processing programs are designed to help writers generate and use text efficiently. We've discussed the core editing commands in word processors elsewhere. In Chapter 1, we introduced you to the CUT, COPY, DELETE, and PASTE commands, as well as the FIND and REPLACE commands. In Chapter 2, we extended our discussion of FIND and REPLACE to their use in text editing. In this chapter, we focus on two additional sets of commands that can greatly speed your ability to write documents: moving text and navigating text.

Moving Text Efficiently. One of the major advantages of word processing programs over paper and pencil or typewriters is the ability to quickly move text from one location to another within a document. Moving text entails selecting the text, either with the mouse or with keystrokes, then using one of four techniques to move the selected text to a new location. Although the specific keystrokes and menu commands can vary among word processing programs, the fundamentals remain the same:

- Using the mouse, the keyboard, or menu commands to select text
- Cutting, copying, or dragging text
- Moving the text to a new location

Which technique you use depends upon your preference and your skills using the different techniques. Try each technique to learn which one helps you speed your writing the most.

Selecting Text. Word processing programs allow you to select text in several ways. You can select text with a mouse by clicking at the beginning of the text you want to select and dragging the mouse (moving the mouse while holding down the button) until you reach the end of the text you want to select. You can also use the mouse to select text by clicking on words, sentences, and paragraphs. Often, two clicks will select a word and a third will select a sentence or paragraph. You can use the mouse in a third way to select text: in some programs, holding down the control key and clicking with the mouse in the margin will select a paragraph or even an entire document.

Many word processing programs also allow you to select text via menu commands. A typical menu command would involve clicking on the Edit menu, then clicking on SELECT . . . or SELECT ALL.

Finally, you can select text using the keyboard. Typically, you can hold down the shift key and move the cursor using the arrow keys, the page up or page down keys, and the end or home keys.

Cutting, Copying, and Dragging Text. Once you've selected text, you're ready to move it to a new location. Typically, writers cut or copy text using menu commands, a button bar, a right mouse menu, or keystroke commands. We discussed the merits of each of these techniques earlier in this chapter. In addition to these techniques, you can also move selected text by dragging it with a mouse. Dragging —or moving your mouse as you hold the mouse button down—is a technique that has many advantages. It saves time, and for many writers it's more intuitive than cutting and copying text. To drag text from one location to another:

1. Select the text
2. Place the cursor on the selected text
3. Click on the left mouse button
4. Drag the text to the desired location
5. Release the left mouse button

In some word processing programs, you can copy text from one location to another by holding down the control key as you drag the text.

tip 6.3 ■ Use Automatic Correction Tools

If you're a typical typist, you mistype some words. As word processing programs have become more sophisticated, they've introduced automatic correction tools. These tools include a list of common typing errors or misspellings, spacing problems, and other corrections. If you make a recognized error, the word processing program automatically corrects your error as you write.

You can also add more words to the automatic correction tool. To add words:

1. Click on the Tools menu
2. Click on the name of the automatic correction tool
3. In the dialogue box, type the incorrect spelling in the Replace window
4. Type the correct spelling in the With window
5. Click on **ADD ENTRY**

Beyond making spelling changes, explore the options available in your automatic text correction tools.

Navigating Longer Documents. As you write on the computer, your words disappear, and you'll quickly find the need to move around within your document. Word processors and computers provide a range of commands to help you move through documents. The techniques include using:

- The Go To dialogue box
- Keyboard commands
- Scroll bars

Navigating through a document requires that you have some idea of where you are currently located. To determine your location, you can check the position of the button in the vertical scroll bar to the right side of the screen or you can check the page number displayed in the status bar, usually located at the bottom of the screen.

The Go To Dialogue Box. When you're working on longer documents, the Go To dialogue box allows you to move easily from one page to another (Figure 6.6). In most word processing programs, you can access the Go To dialogue box via menu commands (EDIT then GO TO) and by clicking on the status bar at the bottom of the screen. Before you use the GO TO command, note what page you are on should you want to return to it. To access GO TO from the menu:

1. Click on the Edit menu
2. Click on GO TO
3. Type the number of the page you want to move to, or select another option
4. Click on OK

The program should jump quickly to the desired page in your document. Once you're on the page, you can scroll up or down to reach the desired lines.

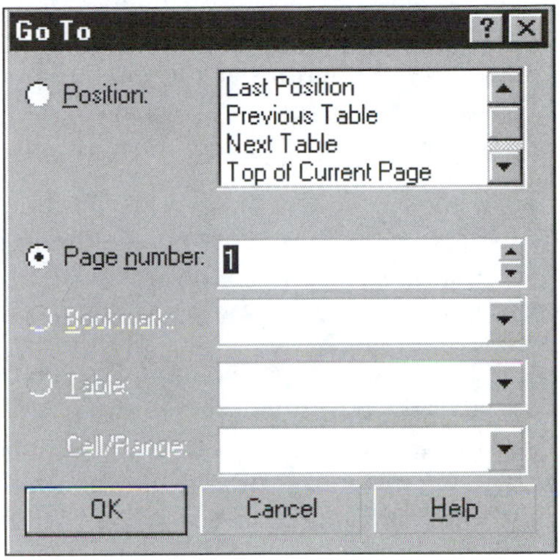

FIGURE 6.6 The Go To dialogue box

Keyboard Commands. You can also move through a document using the keyboard. You can use the arrow keys, page up and page down keys, the home and end keys, and each of these keys in combination with the control key. Depending on your word processing program, other keystroke commands may be available as well. Consult your word processor's manual or online help for more information.

Common keystrokes and keyboard commands for moving through a document include:

- Up Arrow: Move one line up
- Down Arrow: Move one line down
- Right Arrow: Move one space right
- Left Arrow: Move one space left
- Page Up: Move up one screen (roughly half a page, or 18 lines)
- Page Down: Move down one screen (roughly half a page, or 18 lines)
- Home: Move to the beginning of the current line
- End: Move to the end of the current line
- Ctrl + Up Arrow: Move to the beginning of the current line or paragraph
- Ctrl + Down Arrow: Move to the beginning of the next line or paragraph
- Ctrl + Right Arrow: Move one word to the right
- Ctrl + Left Arrow: Move one word to the left

- Ctrl + Home: Move to the beginning of the document
- Ctrl + Home: Move to the end of the document

Scroll Bars. Most word processing programs include a vertical scroll bar on the right hand side of the window. If you are working with a wide document, you can also choose to display a horizontal scroll bar. You can use the scroll bar to move through the document in three ways: by dragging the button in the scroll bar, by clicking on the buttons at the top and bottom of the scroll bar to move up or down one line at a time, and by clicking within the scroll bar to move up or down one screen at a time. To drag the button in a scroll bar:

1. Move the mouse cursor to the button in the scroll bar
2. Click on the button with the left mouse button
3. Hold the left mouse button down
4. Drag the button up or down in the scroll bar by moving your mouse

As you move the button up or down, you move quickly through the manuscript. Note that the position of the button in the right scroll bar signals your approximate location in the manuscript. In leading word processing programs, you will also notice that a flag will pop up when you click on the scroll bar, indicating the page number and section as you scroll through the document (Figure 6.7).

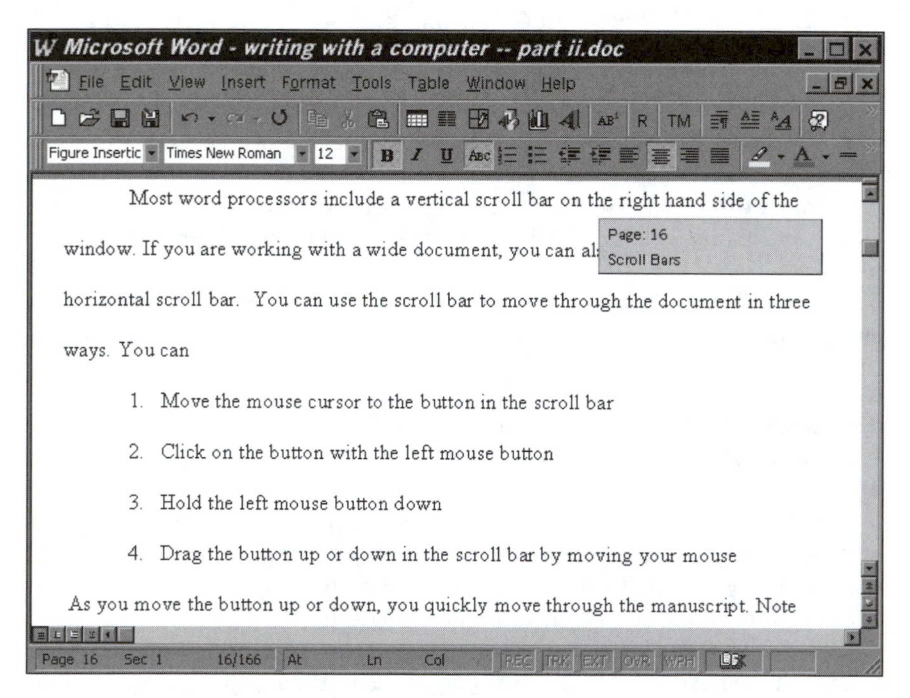

FIGURE 6.7 Page number and section displayed in the vertical scroll bar

tip 6.4 ■ Draft Single Space, Print Double Space

Many publication style guides call for double-spaced text rather than single-spaced text. When that's the publication style, write your documents single-spaced on screen, then move to the top of the document, use the appropriate editing commands to select the complete document, and reset line spacing to double-spacing. The specifics vary with different word processing programs, but they function in a similar way.

When you write a single-spaced document and plan to convert it to a double-spaced document, do not place carriage returns between the heads, paragraphs, illustrations and other graphical elements. If you do and you convert the file to a double-spaced document, you'll have unwanted spacing between the paragraphs, heads, and illustrations. To illustrate:

1. Move your cursor to the top of the file
2. Click on the Edit menu
3. Click on **SELECT**
4. Click on **ALL**
5. Click on the Format menu
6. Click on **LINE**
7. Click on **SPACING**
8. Set the spacing in the Line Spacing dialogue box

The word processing program should, depending on the size of your document and the computer speed, double space your document within a few seconds. After the program finishes double spacing, use the page up and down keys to move through the manuscript checking for any extra lines between paragraphs, heads, illustrations, and other elements.

Customizing Your Word Processing Program

Designers of word processors work extremely hard to design programs that can be used easily by a large number of writers. They create menus that contain commands that most people seem to need. They design button bars with icons that seem to be understood by the majority of the people on whom they've tested the program. And they set the default options on writing tools, such as the spelling checker and the grammar and style checker, in ways that seem to meet the needs of most people.

If you're perfectly average—that is, if you're just like most people—you won't need to customize your word processing program. Frankly, however, we doubt that anyone finds everything to their liking in even the most popular programs. Instead, most writers decide that they can live with the defaults. As a result, they invest little if any time setting up their word processing programs to reflect their individual needs.

If you're not content to live with a program that's configured for the average writer, you'll find this section useful. We focus here on common customization options, paying particular attention to the following changes that can help you write more effectively and efficiently: customizing basic options, colors and fonts, keyboard commands, menus, button bars, macro commands, and writing tools.

Customizing Basic Options. Basic options are settings that change the overall appearance and function of your word processing programs. In most programs, you can access the Options, Preferences, or Settings dialogue boxes through the Edit or Tools menu. The dialogue box allows you to set options such as how you view documents, how you edit documents, and how you save and print documents (Figure 6.8).

The basic options dialogue box lets you make numerous changes in how you use your word processor. In one widely used word processing program, Microsoft Word, more than 125 options can be set through the options dialogue box. These options include:

- Whether clicking and dragging your mouse across a word automatically selects the entire word
- Default locations for saving and opening files
- Whether to confirm file conversions when opening files saved in another word processing format
- Whether to view white text against a blue background or black text against a white background
- Whether to view scrollbars
- Whether to show nonprinting characters (such as spaces, tabs, and ends of lines)

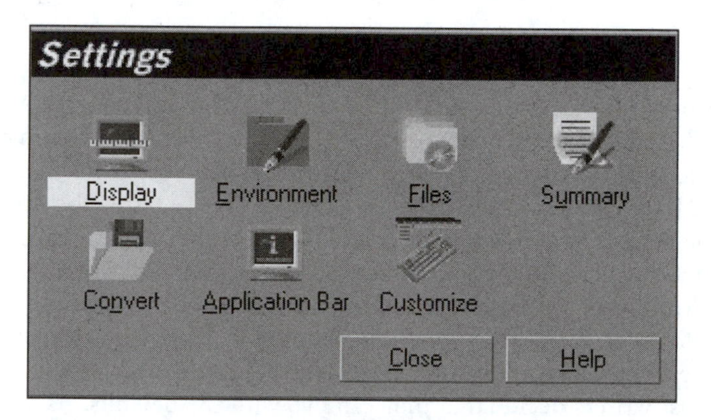

FIGURE 6.8 The Settings dialogue box

It's unlikely that you'll want to change more than a small minority of these options. However, making even small changes in the appearance of your word processing program can greatly speed your work. For instance, if you are working with text that contains numerous page breaks in unexpected places (as you might find had you copied text from a Web page or an electronic mail message), you might find it easier to edit the document if you could view nonprinting characters, such as spaces and hard returns. Similarly, many writers find it frustrating to work with automatic word selection. Using the Options dialogue box, you can easily change these and other settings to more accurately reflect your needs as a writer.

Customizing Colors and Fonts. You can customize colors and fonts in most word processing programs in two ways: by modifying the template that you use when you open a new document and by modifying the format of your default paragraph style. Since styles are found in templates, modifying a template or modifying a style amounts to essentially the same thing, but only if you remember to save the style as a global setting—that is, if you save it to the document template that you use for most of your documents. Typically, this template is called the *normal*, *default*, or *document* template. If you don't save the style to a template, it will change only the appearance of the current document.

Black text on a white background works well for most writers. Since this is the default setting for most word processing programs, few writers change it. Depending on the lighting in the room where you use your computer, however, you may find it more effective to specify a different, possibly a darker, background color.

Some writers find that the default font size and face in their word processing programs—typically 12-point Times New Roman—isn't to their liking. Some prefer to set the default font to a larger size and different face, such as 12-point Courier or 13-point New Century Schoolbook. Depending on the resolution at which you view your desktop and your eyesight, you may prefer to change the default font setting.

Customizing Keyboard Commands. Keyboard commands refer to the keys you push to invoke a particular command. Since your hands are usually on the keyboard as you type, keystroke commands are typically the fastest way to tell the word processing program to do something. For instance, it's much quicker to type Ctrl-b to format selected text as bold than it is to move your hands from the keyboard to the mouse and then to use either the button bar icon or menu commands.

Word processing programs come with a set of default keystroke commands, but each command can be customized (Figure 6.9). For example, in one leading program, the default keystroke command for FIND AND REPLACE is Ctrl-h. Many people find this far from intuitive. If you've grown accustomed to the keystroke command used for FIND AND REPLACE in other word programs, such as Ctrl-f, you might want to change the keyboard command from Ctrl-h to Ctrl-f. If you

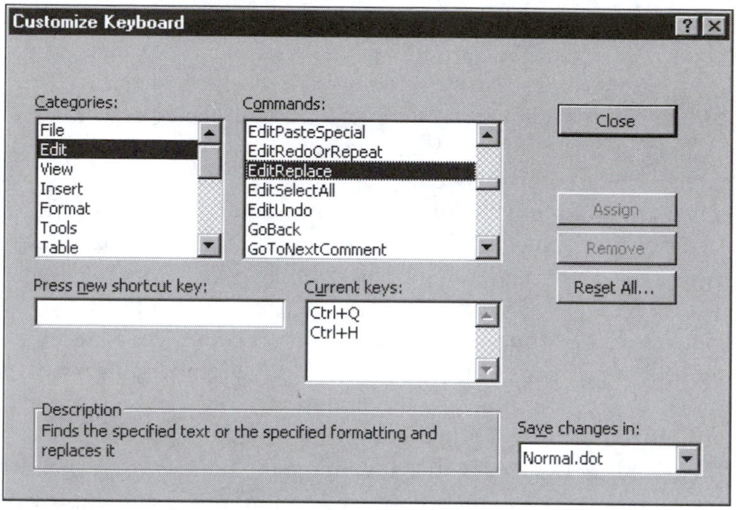

FIGURE 6.9 The Customize Keyboard dialogue box

don't like either Ctrl-h or Ctrl-f, you can use a different keystroke combination, such as Ctrl-r or Alt-r.

Before changing a keystroke command, consider possible complications. Often, the combination of keystrokes that you want to use is already assigned to another command. Ctrl-r, for instance, might be assigned to the RIGHT JUSTIFY command. The Customize Keystroke dialogue box will let you know what the keystroke combination you'd like to use is assigned to. Once you know what the assignment is, you can decide whether to change it.

Customizing Menus. You can also customize your menus. You can specify which commands appear on specific menus and their order. For instance, you might want to simplify your menus so that fewer commands are available, thus making it easier to locate the commands that you regularly use. Removing a command from a menu is quite simple in most leading word processing programs: open the Customize, Settings, or Preferences dialogue box by right clicking on the menu using your mouse or by using menu commands, then open a menu and drag items off of it (Figure 6.10).

Adding items to a menu is equally straightforward. In the Customize dialogue box, locate the command that you want to add to a menu, then drag it to the menu and place it where you want it to appear.

Customizing Button Bars. Button bars provide one-click access to commands such as NEW FILE, OPEN FILE, SAVE FILE and PRINT. Leading word processing

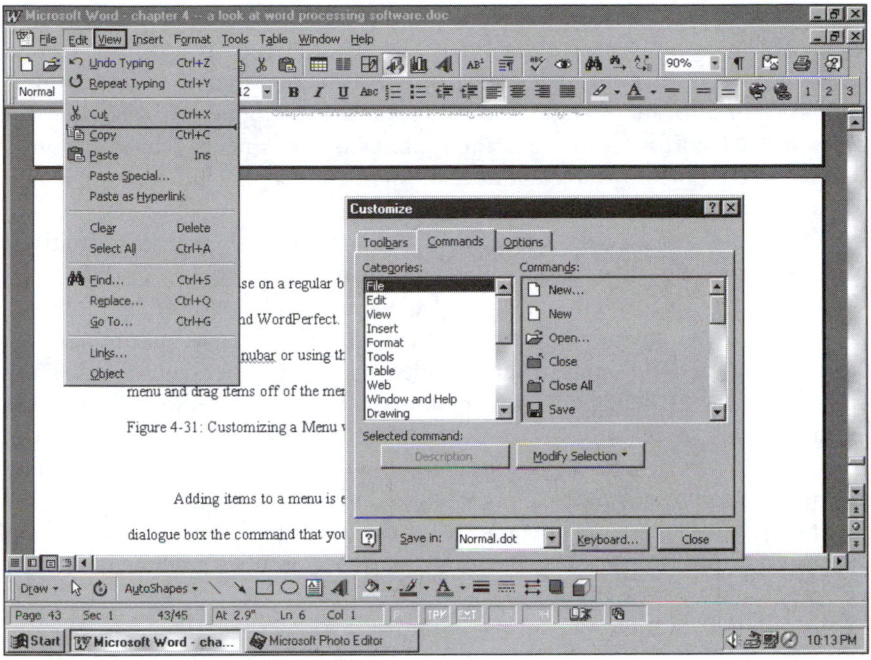

FIGURE 6.10 Customizing a menu

programs provide several standard button bars, some for general formatting of documents, others for working with tables or borders, and still others for working with pictures or forms. You can customize which button bars appear when you open a new document, where they appear (along the top, bottom, or side of the screen), and what commands they display. You can also create or delete button bars.

You customize button bars in the same way that you customize menus. You drag icons off of button bars, add new icons to button bars, and change the order in which icons are displayed. You can display commands on a button bar using text, icons, or a combination of text and icons.

When the designers of word processing software create standard button bars, menus, and keyboard commands, they try to accommodate the needs of a wide audience. In contrast, you only have to satisfy yourself. By customizing your button bars, menus, and keyboard commands, you can provide easier access to the commands you use most frequently. Doing so can increase your writing efficiency.

Creating Macro Commands. One of the most powerful tools available to writers are macros, or sequences of commands that can be recorded and subsequently repeated as needed. A macro can be quite simple: for example, inserting symbols such as an em dash (—) into a document. Using menu commands, this can be a time-consuming process. For instance, you can click on the insert or text menu, select **SYMBOLS**, locate the em dash in the symbols dialogue box, and then click on the **INSERT** button. To reduce the number of steps needed to insert an em dash, you can record a macro that does it for you and assign an icon to your button bar to run the macro with a single mouse click.

Macros can also be quite complicated. You might find yourself formatting a series of documents that must be changed in numerous ways. These changes might involve a series of **FIND AND REPLACE** commands followed by global format changes to the document. Using a macro, you could record all of these commands as you carried them out on a single document. Then you could run the macro on the other documents, saving yourself time and a tedious repetition of commands.

Leading word processing programs allow you to create, edit, and assign macros to button bars, keystroke commands, and menus. Although learning how to work with macros can take some time, it's a sound long-term investment. If you regularly invoke sequences of commands as you write, using macros can significantly reduce your writing time.

Customizing Writing Tools. Writing tools such as Spelling Checkers, Grammar and Style Checkers, and Autocorrect can also be customized. In addition to adding words to a spelling checker, you can change settings such as whether to check your spelling as you type, which dictionary to use, and whether to suggest

Panel 6.2
Sorting

Sorting works especially well for organizing lists and references. When drafting text, you don't need to worry about putting your entries in correct order. Instead, you can type the information and use the computer to sort the items. Computers can sort alphabetically or numerically in either ascending order (A to Z, 0 to 9) or descending order (Z to A, 9 to 0).

Before you sort large lists, make backup copies of your file. If something goes wrong, you can always return to your original list. Be especially cautious when sorting paragraphs and references. Before you use the **SORT** command, review the guidance on sorting in the online help. Read it carefully. If you like, print a copy and have it in front of you as you try the **SORT** command.

Assume you're working on a document, and you have created a reference list (sometimes called References Cited or Bibliography depending on the academic field). Assume further that you entered each reference as you drafted your document, and changes when flagging a potentially misspelled word. Similarly, you can specify whether to check grammar and style as you type and, if so, which grammar and style rules to use.

The location of settings dialogue boxes for writing tools varies widely from word processing program to program as do the options that you can set. In Microsoft Word, you'll find the spelling and grammar settings in the Options dialogue box, while the settings for Autocorrect are found in the Tools menu.

Using Online Help

A word processing program's built-in online help can be a powerful resource—and a great time saver. Although it's tempting to hunt through the menus looking for a particular function, such as how to insert a graphic into a document, consulting online help can return you to writing more quickly than using your best guesses to figure out how to use your program.

Online help comes in several forms (Figure 6.11A–C). In most word processing programs, you can obtain help through the Help menu. In many programs, you can also obtain help through context sensitive suggestions in dialogue boxes. In leading word processing programs, you can also obtain help through an "assistant," "guide," or "expert." This form of help waits in the background until you appear to be trying something new. At that point, it will pop onto your screen and ask if you want help. You can type questions about topics and the expert will offer its best guesses at appropriate information. Some writers find experts intrusive. If you do, you can turn off the expert and seek help in other ways.

Panel 6.2 *(continued)*

they are not in alphabetical order. You need to alphabetize them. Before you begin, make sure you have a hard carriage return at the end of each reference—in other words, each one is a separate paragraph. To sort the list:

1. Highlight the list
2. Click on the Tools, Text, or Table menu
3. Click on **SORT** and review the dialogue box
4. Determine whether the default is set for paragraphs
5. Note the options, such as **ALLOW UNDO AFTER SORTING**
6. Click **OK** to begin the sort

The program should automatically sort your list. If the sort does not perform the way you envisioned, you can undo the sort as follows:

1. Click on the Edit menu
2. Click on **UNDO**

Help Topics: Microsoft Word

Contents | Index | Find

Click a book, and then click Open. Or click another tab, such as Index.

Key Information
Getting Help
Installing and Removing Word
Running Programs and Managing Files
Opening, Creating, and Saving Documents
Typing, Navigating, and Selecting
Editing and Sorting
Checking Spelling and Grammar
Formatting
Changing the Appearance of Your Page
Importing Graphics and Creating Drawing Objects
Working with Tables and Adding Borders
Working with Long Documents
Sharing Data with Other Users and Applications
Working with Online and Internet Documents

Open | Print... | Cancel

FIGURE 6.11A Online Help dialogue box

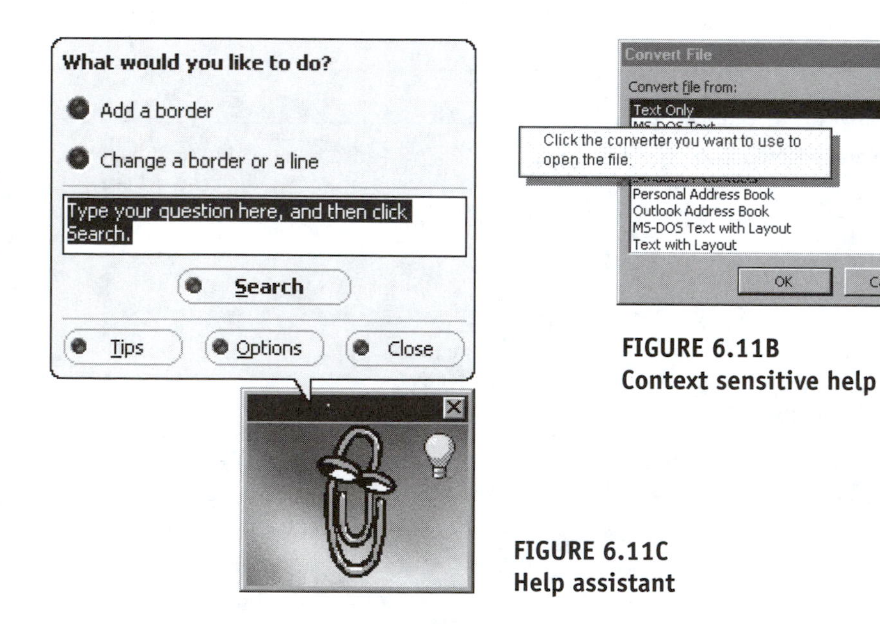

FIGURE 6.11B Context sensitive help

FIGURE 6.11C Help assistant

As word processing programs have grown more sophisticated, the amount and variety of online help has increased—as have the ways in which help can be obtained. You can obtain information by:

- Consulting a table of contents or an index
- Searching for help on a topic
- Using a "wizard" that leads you through the steps in a particular task, often using demonstrations
- Consulting company Web sites for additional information, advice, and options for obtaining feedback on specific issues (either from company representatives or from other users of the program)

REUSING DOCUMENTS

Word processing programs provide the opportunity to recycle and reuse documents—techniques that can save a significant amount of time. Do not recycle or reuse documents from others without their permission and without giving them full credit (Panel 6.4). In this section, we focus on recycling documents, recycling sections of documents, and creating boilerplate for use in multiple documents.

Recycling Documents

If you're a student, you may be asked to write a series of documents that culminate in a major project. A term paper, for example, may begin with a formal proposal, continue with a series of progress reports, and near completion with a rough draft before the final term paper is submitted. Similarly, if you're a professional, you may often be asked to write proposals, progress and final reports on a wide range of projects. Word processing programs come in especially handy for writing a series of documents reporting on a project, as you can recycle all or part of earlier documents when writing later documents.

Assume you've written a proposal for a term paper. Your proposal includes a brief literature review, an outline for your term paper, a list of the references you cited in your initial literature review, and a timetable for completing the term paper. You've named the proposal file *prop.doc*. To write the progress report, open *prop.doc*, use the SAVE AS command and rename it *prog1.doc* (for progress report number one). Add, delete, and revise the text to reflect your progress on the project. The specifics may vary for your progress report, but you should include what you have accomplished and what you have left to accomplish.

When you begin writing the rough draft of your report, you can use the last progress report or the original proposal as a foundation. Assume you open the *prop.doc*, then rename it as *roughdraft1.doc* (for the first draft of your rough draft). You can expand your text and elaborate on your research. As you write, you can add text under the different sections of your outline. You'll probably find some sections may go quicker than others. If so, jump from section to section within a document and do not worry about writing the document from front to end. Simply move to a section and begin writing. When you exhaust your information or tire of writing that section, jump to another section and begin drafting additional text. (Before you jump to another section, save your document and make a backup copy to ensure that you don't lose what you've written.)

When some writers save their draft documents, they save them as a series of files, for example, *reptdf1.doc* for draft 1, *reptdf2.doc* for draft 2, *reptdf3.doc* for draft 3, and so on. In this way, should they delete sections or decide that an earlier organization works better after all, they won't need to rewrite the entire document. Instead, they can open an earlier version of the document and revise it as needed or copy sections from an earlier draft into a later draft.

Recycling Sections of Documents

Word processing programs enable you to open two or more documents and to copy text in one document and paste it in another document. Recycling text enables you to speed write. As we were drafting this book, we used this strategy for several chapters. We'd written a prospectus (a proposal) for this book, and submitted it to our publisher. Once we signed a contract, we used the recycle method to begin each chapter. We copied the chapter summaries from our prospectus into several new documents, using each summary as the starting point for a chapter.

To illustrate the specific commands, consider our method for this chapter. First, we:

1. Clicked on the File menu
2. Clicked on OPEN
3. Located the file on our hard drive
4. Clicked on *prospectus.doc*, the book's prospectus

With the book prospectus opened, we:

1. Clicked on the Edit menu
2. Clicked on GO TO
3. Entered 5, the page number where the chapter's summaries began, in the Go To dialogue box
4. Used the page down and arrow keys to locate this chapter number, title, and summary

Next, we:

1. Selected the text containing the chapter number, title, and summary
2. Copied the selected text
3. Clicked on the File menu

4. Clicked on **NEW**
5. Selected the template
6. Clicked on **OK**

With the new file open, we:

1. Clicked on the Paste icon in the button bar
2. Began outlining and writing the chapter

You can also use this recycling strategy to move sections from an existing document to another. To illustrate, assume you were a Colorado State University student and you were living on campus in the summer of 1997. The thunderstorm moved through Fort Collins on the evening of July 28. In about four hours the thunderstorm dropped more than 9 inches of rain just west of campus. A flood moved across campus causing damage of roughly $100 million to more than two dozen buildings. That sparked your interest in disaster prevention, recovery, and a dozen related topics. As opportunities arise in different classes, you write a report about students' reactions to the flood for a psychology class. You name the final document *Student Reactions to Flood.doc.* To set the stage for discussing student reactions, you write a two page description of the meteorological accounts of rainfall, sequence of flooding on campus, and a brief description of the damages.

Later, you take a basic economics class, and you're asked to write a term paper on an economics topic. Your observations of the Colorado State University clean-up and disaster recovery efforts stoke your interest in the economics of disaster recovery. You research the topic and begin drafting your term paper, which you save as a file named *Economics of Flood1.doc.* As you work on *Economics of Flood1.doc*, you decide that adding a brief account of the Colorado State flood will help illustrate the local events. You begin writing an account of the flood and realize you could recycle that section from your *Student Reactions to Flood.doc.* Here's how you can move the three paragraph description of the flood from *Student Reactions to Flood.doc* to *Economics of Flood1.doc.*

To move text from one document to the other:

1. Click on the File menu
2. Click on **OPEN**
3. Locate the *Student Reactions to Flood.doc* on your computer
4. Select it and click **OK**

Next open your document, *Economics of Flood1.doc:*

1. Click on the File menu
2. Click on **OPEN**
3. Locate the *Economics of Flood1.doc* on your computer
4. Select it and click **OK**

You now have two files open. In many word processing programs, you can work with the open files in different ways. You can toggle between the two files, or you can arrange the files in windows so that you can view small portions of each document at once on the screen. Toggling between two open files provides the largest viewing area of the different techniques. To illustrate:

1. Click on the Window menu
2. Note the menu items, such as **NEW WINDOW**, **ARRANGE ALL**, **SPLIT**, **CASCADE**, and **TILE**
3. Note, below the options, the file names
4. Click on *Student Reactions to Flood.doc*
5. Scroll to the section providing your historical account of the flood
6. Select the historical account of the flood
7. Click on the right mouse button
8. Click on **COPY**

After you've copied the historical section, switch to your document, *Economics of Flood1.doc* about the economic impact of the flood, and paste the three paragraph historical section into the economics draft. Use the following commands:

1. Click on the Window menu
2. Click on *Economics of Flood1.doc*

When it opens up:

1. Move to the insertion point
2. Click on the right mouse button
3. Click on **PASTE**

The program will insert the historical account into your new document. Although the basic process remains the same, the menu and other commands may differ among programs.

Return to the Windows menu and note the selections. You can view each window as a split screen, or you can tile or cascade the windows. To read more about viewing windows, see Panel 1.2.

Creating Boilerplate

Boilerplate refers to text you can use over and over again in different documents. Using boilerplate is like using macro commands—it allows you to write passages (or even entire documents) once and then use them in a variety of documents. For example, when researchers write proposals submitted to federal agencies, they must submit two page resumes that highlight their careers and recent accomplishments. Rather than rewriting their resumes for each proposal, they produce a master copy, update it annually, and use it repeatedly in different proposals.

If you need to use the same information in different documents, consider creating individual documents containing boilerplate. This is a common practice

Panel 6.3
Integrating Documents

Word processing programs provide the opportunity to merge documents in an almost seamless way. To illustrate, assume you're on a team writing a report on the potential impact of new industries moving into your community. Prior to preparing their final drafts, everyone agrees to save their documents in the same word processing format, and agrees on formatting details and style issues. Assume Charlie wrote a report on economic impacts, *econ.doc;* Sally wrote a section on school impacts, *educ.doc,* Dana wrote a section on air quality impacts, *air.doc,* and you wrote a section on water quality impacts, *water.doc.*

To integrate the reports, you could open the original proposal, *prop.doc,* and rename it, *Industrial Impacts.doc.* To merge the documents:

1. Move to the economics section of the original proposal
2. Click on the Insert menu
3. Click on **FILE**
4. Locate the file in the Insert File dialogue box
5. Select the file, *Econ.doc*
6. Click **OK**

The word processing program will insert the new file into the document at the point where you've placed the cursor. Move to the next section of the report and repeat the sequence, inserting the other files at their respective locations. Continue until you've inserted all files into the master document. After each insertion of a new file into your master document, save the master document again to ensure that you don't lose your work.

in many corporations and agencies. In some cases, they have developed sophisticated software programs that enable them to generate larger proposals quickly by integrating files containing boilerplate into a single document.

Boilerplate provides a way of speeding your letter writing. Assume you're searching for a job. You can create a series of different paragraphs as boilerplate so that you need not write each letter from scratch. Instead, you can select from the various paragraphs, insert their respective files into your letter, and then revise as needed to tailor the letter to fit the specific job.

LOOKING AHEAD

Almost without exception, writing projects take more time than we expect. Few writers, as a result, would consider using a computer if they knew it would slow them down. But a surprisingly large number of writers make little or no use of time-saving commands, tools, and writing strategies made possible by the com-

Panel 6.4
Ethical and Legal Considerations

Easy access to electronic copies of documents has made it easy to use parts of or even entire documents written by other writers. In some cases, plagiarism—using another person's work without giving due credit to the author—is unintentional. In others, it's deliberate. Regardless of intent, you should avoid plagiarism. Plagiarizing another writer's work is unethical and illegal. In academic and business settings, writers who plagiarize face stiff penalties. Students can fail a course or even be expelled; professional writers can lose their jobs or face legal sanctions.

Although some people plagiarize without being caught, others have ruined their academic and professional careers. Consider two roommates, Mary and Ann. One day while Ann is in class, Mary turns on Ann's computer and opens a file containing an essay Ann had written for a composition class. Mary types her name in place of Ann's, prints the document, and turns the essay in to her writing instructor. The instructor recognizes Mary's latest paper as having a different writing style than her other work. With some digging, the instructor learns that Mary plagiarized the paper. The instructor gives Mary a "0" for the paper. While schools, colleges, and universities vary on academic dishonesty policies, students may receive an "F" or "0" on the assignment, fail the course, or be dismissed from school.

Consider Sam, an editor of company X's magazine. A year earlier, Sam had written an especially creative article and published it in his company magazine. Imagine his surprise when he read a magazine from company Y and saw his original article with Ed Carpenter's byline. No credit was given to Sam—Ed had plagiarized his work. Sam was livid. He picked up the telephone, found Ed's telephone number, and called him. Sam identified himself as the editor of company X's magazine and told Ed he had found the article especially well written and interesting. Ed basked in Sam's praise, failing to recognize Sam as the original author. Rather than accusing Ed of plagiarism, Sam asked Ed for permission to reprint the article in Sam's company's magazine. In response, Ed sends a letter giving Sam permission to reprint Ed's article (Sam's original article). Armed with copies of Ed's plagiarized article, Ed's permission letter, and Sam's original article, Sam contacted the president of Ed's company. The embarrassed company president apologized and asked for copies of the articles and letter. Shortly after Sam sent them, Ed was fired.

In the second case, Sam had not only crossed an ethical line, he had stepped into deep legal waters. Sam's company could have sued Ed's company for violation of copyright law. The 1976 copyright law protects the rights of authors, artists, and other people who produce creative work—writers, illustrators, photographers, musicians, multimedia developers, and others. Basically, copyright law says that authors, artists or other creators should receive financial remuneration for use of their works. If you work for a company and produce creative works as part of your job, the works belong to the company—such works are considered "works for hire" under the copyright law. When someone uses those works improperly, the author or owner can bring a lawsuit. Fines and damages can be charged against the violator.

Panel 6.4 *(continued)*

The copyright law provides for using another's work under three conditions: fair comment, fair use, and permission. Fair comment usually entails making an editorial comment on the quality of a document, such as occurs in a book review. In the review, reviewers can quote from a passage without permission. Fair use enters a murky legal abyss, depending on the length and nature of the original work. Generally, you can't use short works, such as poems, art work, photographs, data tables, and similar materials without permission. Major book publishers provide "fair use" guidelines for authors of new books and specify that authors cannot use more than a total of 300 words of text from another work without written permission. The 300 words can be taken from one or more places in the work. If however, the 300 words are from a short text such as Martin Luther King Jr.'s "I Have a Dream" speech, then you would need to seek permission from the author or copyright holder, in this case Coretta Scott King, the late Martin Luther King Jr.'s widow.

Recycling documents raises more than issues of plagiarism and copyright infringement. You also need to consider the ethics of recycling your own materials for different documents. If you're a student, and you've written a term paper for one course, you might be tempted to recycle the same paper for another course. Think twice, however. Many instructors do not accept such practices. It may be acceptable, however, to use the same research and background information to develop a new term paper with a different focus for another course.

Similarly, researchers who have written an article for one academic journal should not recycle the same article to another journal without discussing its prior publication with the editor of the second journal. Most academic journals require that the articles they accept have not been published before. However, it's an acceptable practice for an author to present a paper at a conference, revise the article, and then submit it to a journal. When authors use such an approach, they acknowledge that they gave a presentation or paper at a conference when submitting the article to a journal.

Magazines often request first publication rights—in other words, the magazine accepts only manuscripts for articles that have not been published previously in magazines. Exceptions do arise, however. If the article would be of interest to the readers of different magazines, authors can query editors of different publications asking if they are interested in the article. If they are, the article can be printed along with an acknowledgement that it had been printed in an earlier publication. Many religious magazines, for example, have narrow audiences and their readerships do not overlap. In such cases, editors may accept articles published in other magazines. The key criteria is nonoverlapping readership.

puter. As you look ahead to your next writing project, consider this chapter. If you incorporate only a few of the commands, tools, and strategies we discuss, you'll still save plenty of time. If you take the time to learn most of them, you'll significantly decrease the time it takes to complete your projects.

chapter 7

Organizing Your Computer's Desktop

Most computers have a desktop—a graphical representation of the files and programs available on the computer. In more technical terms, a desktop is part of a computer operating system's graphical user interface (or GUI—"gooee"). The basic principle of a desktop, and for that matter a GUI, is fairly simple: icons, small drawings that represent files and programs on a computer, can be moved around the desktop so that it's easier to use the computer.

"Easier to use" is a relative idea. The GUI is generally seen as an improvement over earlier computer operating systems that used command lines. Command line operating systems—such as DOS or UNIX—use typed commands to tell the computer which programs to start or what to do with specific files. Memorizing these commands takes time—typically more time than it takes to learn how to use operating systems that use graphical interfaces (Figure 7.1A–D).

The GUI dates back to the 1970s when it was developed by researchers at Xerox Corporation's Palo Alto Research Center. In the 1980s, Apple introduced the first commercial GUI, the Macintosh Operating System. Microsoft followed with several generations of its Windows operating systems, IBM introduced the OS/2 system, and several flavors of UNIX adopted a graphical user interface. Today, few computers use command line operating systems.

But don't worry, the designers of most GUI-based operating systems haven't taken all the mystery out of using a computer. Plenty remains to learn about working with what you see on your computer screen. In this chapter, we'll talk about setting up your computer's desktop to help you write better. We'll focus on the desktop metaphor, the makeup of your desktop, and how to set up a desktop that works for you.

```
 Volume in drive D is Texts and Data
 Volume Serial Number is D446-8653

 Directory of D:\BIBS

04/08/98  01:54p        <DIR>          .
04/08/98  01:54p        <DIR>          ..
01/13/90  09:45a                16,876 BIB-HIST.DOC
11/05/90  09:33a                23,264 BIB-TEXT.DOC
04/06/95  01:43p               414,208 C&C-BIB.DOC
06/05/92  03:27p                12,383 CANON92.DOC
03/31/92  03:06p                40,350 EXAMBIB.DOC
03/31/92  03:00p                12,115 HISTORY1.DOC
03/31/92  02:57p                 4,902 HISTORY2.DOC
03/31/92  02:59p                 9,032 HISTORY3.DOC
04/01/92  01:12p                18,235 MISC92.DOC
12/04/93  12:36p                 2,477 READINGS.DOC
09/20/95  03:56p             1,610,240 REFMAN.DOC
04/08/98  01:54p        <DIR>          REVIEWS
04/15/92  12:32p                15,252 TOP-TEN.DOC
02/04/97  06:34p                 2,489 WS_FTP.LOG
              16 File(s)      2,181,823 bytes
                          4,849,901,568 bytes free

D:\BIBS>copy c&cbib.doc a:_
```

FIGURE 7.1A DOS desktop

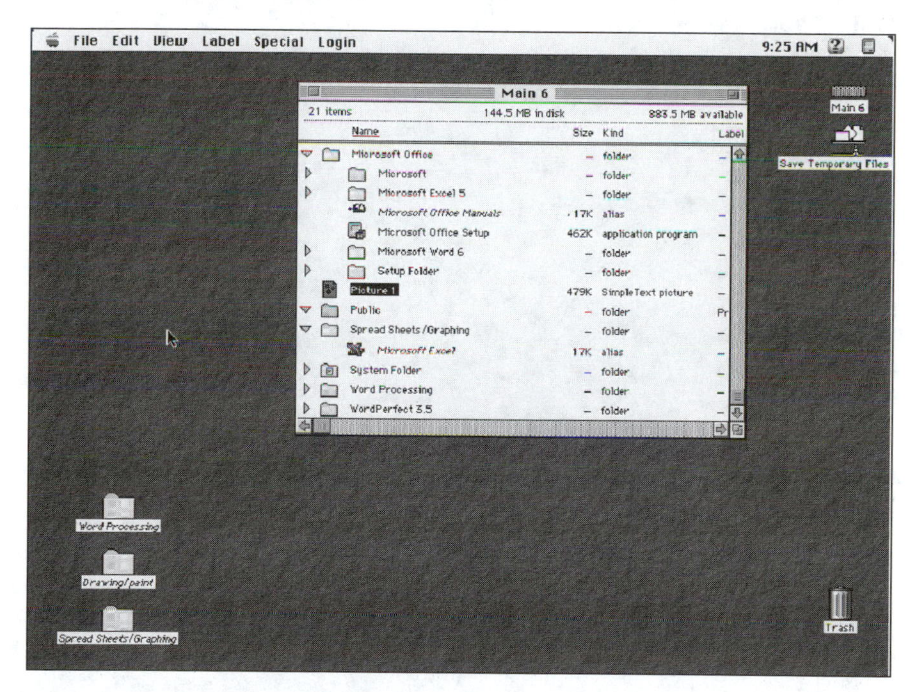

FIGURE 7.1B Macintosh desktop

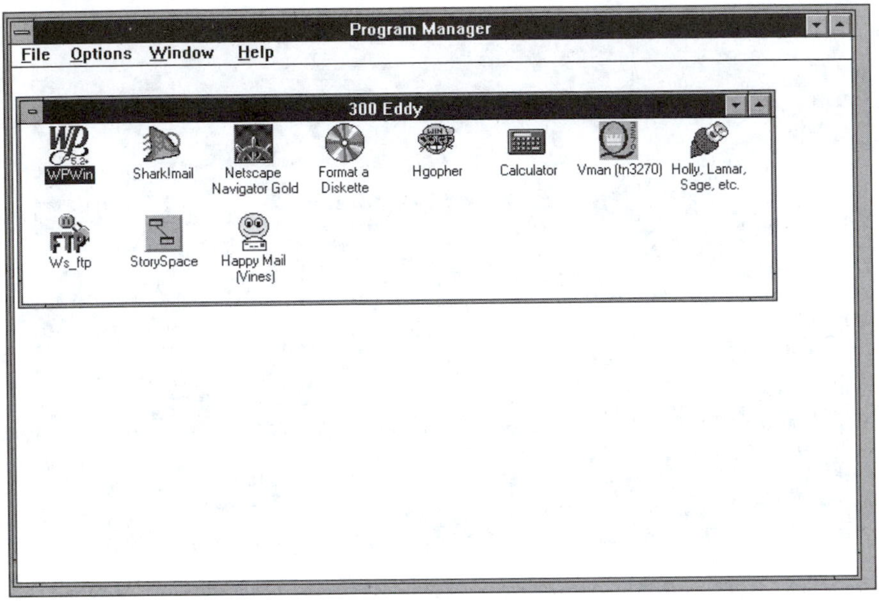

FIGURE 7.1C Windows 3.1x desktop

FIGURE 7.1D Windows 95 desktop

THE "DESKTOP" METAPHOR

The desktop metaphor has become an important part of how people think about computers. It represents things that you'll find on a typical office desk, such as files, file folders, typewriters, calculators, and telephones. As a metaphor, the idea of a desktop works well—if you're familiar with an office, that is. For people who have little or no experience with the metaphor—school children, for instance, or people whose work experiences have been in other areas—the desktop metaphor may not work as well. If you've never seen a file or a file folder, for example, you won't find it intuitively easy to work with the metaphor.

Fortunately, for most people who use computers the desktop metaphor works relatively well. It's common practice to put files into folders, to throw things you no longer need into a trash can or recycle bin, and to organize your desktop so you can find things.

Like almost any metaphor, however, the desktop idea eventually breaks down. Knowing whether to click once or twice on an icon with a mouse—an action characteristic of most operating systems—is far from intuitive. And what's a mouse doing on your desktop, anyway?

If the idea of a desktop is new to you, practice using the system. Start by considering what you'll find on the desktop.

WHAT'S ON YOUR DESKTOP?

Your desktop is made up of icons that represent files and icons that represent folders (also known as directories). Files and folders are the building blocks of computer operating systems. Files can contain text, graphics, sound, video, or programs (such as a word processing program, a graphics editor, or a spreadsheet). Folders contain files and other folders (see Figures 7.2A–E and 7.3).

Think about folders as ways of organizing information. In an office, you put related documents—memos, letters, reports, articles, or stories—into file cabinets. Each file drawer contains a set of hanging file folders. Each hanging file folder contains a group of related files.

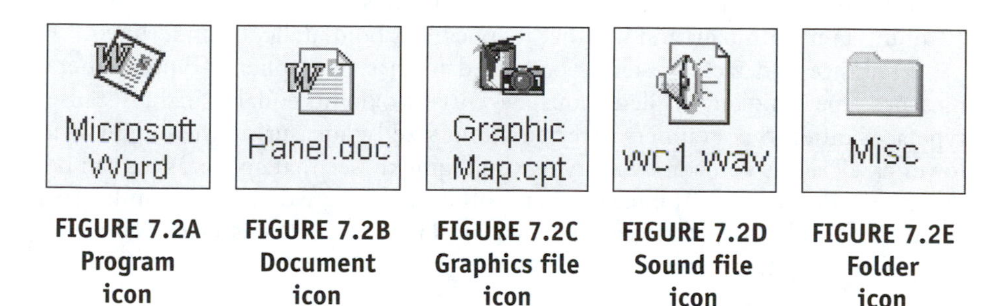

FIGURE 7.2A Program icon

FIGURE 7.2B Document icon

FIGURE 7.2C Graphics file icon

FIGURE 7.2D Sound file icon

FIGURE 7.2E Folder icon

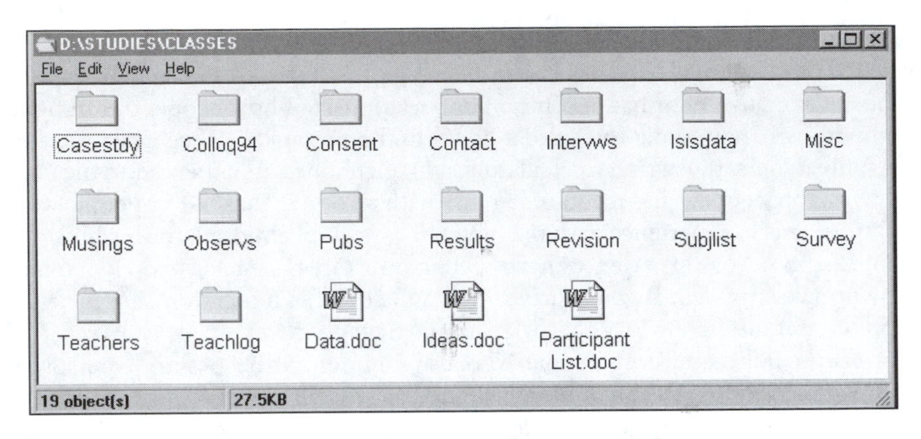

FIGURE 7.3 A folder containing files and other folders

Computer folders and files can be organized in the same fashion. You might create a folder called, *My Novel.* You'd place files related to your novel within the *My Novel* folder. As the number of files in the folder increases, you could decide to organize them further, perhaps by chapter. To do this, you'd create several new folders—*Chapter 1*, *Chapter 2*, *Chapter 3*, and so on. Or you might decide to organize them by function. You could place your chapters in a folder called *Chapters*, your research in a folder titled *Research*, and your letters to publishers in a folder called *Letters* (see Figure 7.4A,B).

Files can vary a great deal as well. Many files will be program files—files that run your computer operating system or particular kinds of software. Other files contain information you create with software programs—such as word processing files and files created by spreadsheet programs. Like folders, you can give files names to help you remember what is in them.

In addition to files and folders, three other elements contribute to what you see on your desktop: fonts, colors, and screen size (or resolution). Fonts refer to the letters and numbers that are used to label files and folders (another use of the desktop metaphor). You can typically specify a font's size and face (for instance, 12-point Times Roman) and whether it appears as bold, italic, or underlined.

Font size and face are terms borrowed from typographers. Typographers measure type using units called points. Seventy-two points equals 1 inch. Because typefaces differ, typographers determine font size by measuring the height of a lower case "x." As you work with type, you'll quickly see that appearances can be deceiving. Because of capitalization and other characteristics, for example, 10-point Arial looks almost as large as 12-point Times New Roman (to read more about fonts, see Chapter 3).

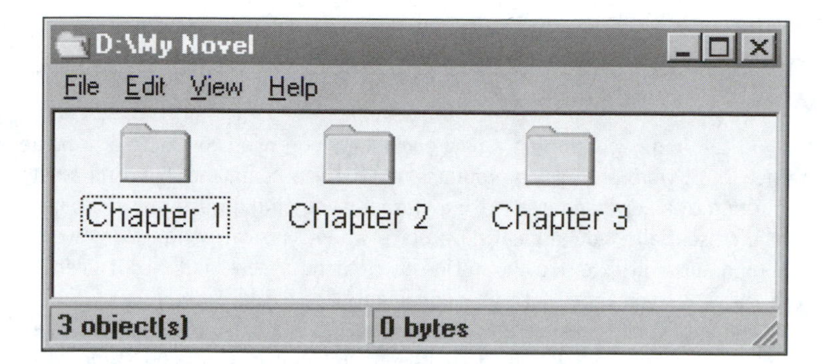

FIGURE 7.4A Folders organized by chapter

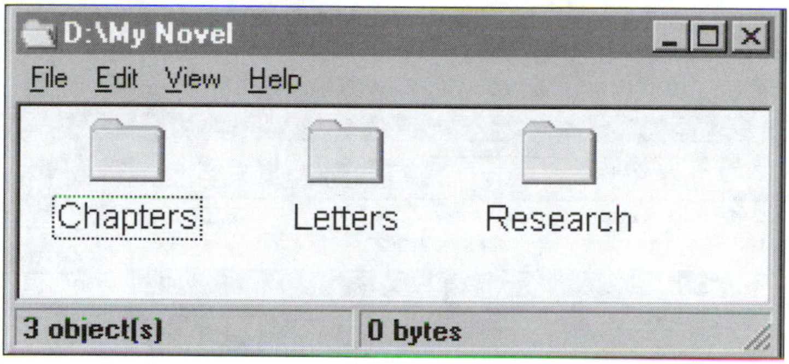

FIGURE 7.4B Folders organized by category

Colors extend the desktop metaphor by bringing in the ability to select a color scheme for your office. In most operating systems, you can even specify whether you want to have wallpaper and, if so, what you'd like it to look like.

Resolution refers to the size of your desktop. On Windows operating systems, for one, you can display your desktop at VGA resolution (640 lines horizontally by 480 lines vertically), Super VGA resolution (800 by 600), or higher resolutions (1,024 × 768, 1,280 × 1,024, or even 1,600 × 1,280). The ideal resolution for your desktop depends on the physical size of your computer monitor and its ability to display higher resolutions. In general, VGA resolution is best for most 14- and 15-inch monitors, while Super VGA and higher resolutions work best on larger (17-inch and up) monitors. If you're working on a laptop with a screen smaller than 10 inches, VGA resolution usually works best.

Panel 7.1
Creating Files and Folders

Before you can begin working on a new document, you need to create a new file. To create new files, word processing programs use the **NEW** command. Creating a new file is like turning to a new page in a writing journal or putting a fresh piece of paper in a typewriter. Essentially, you're creating a place where you can write. Many word processing programs will give you the option of creating a new kind of document when you use the **NEW** command. These new documents are based on templates, which are blank documents with common text and formatting features. Most word processing programs, for example, allow you to create new letters, memos, and standard documents, among a range of other options (to read more about templates, see Panel 3.10).

Occasionally, you'll also want to create a new folder in which to put your files. Many word processing programs allow you to do so with the **CREATE NEW FOLDER** command (Figure A). Often, you'll create a new folder when you're ready to save a new file for the first time.

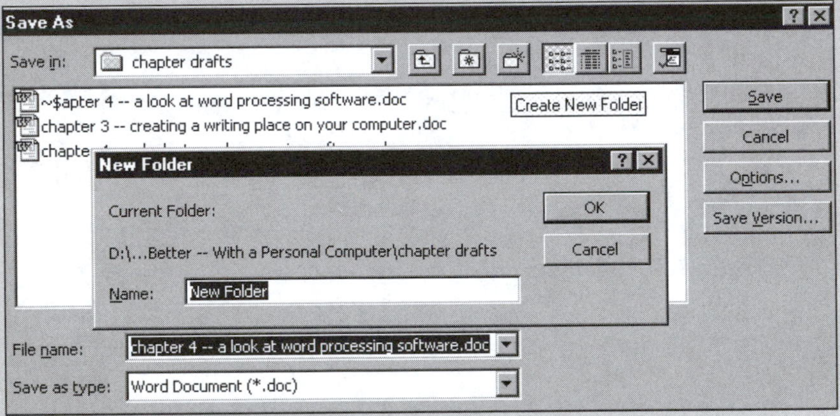

FIGURE A The create new folder command available through the Save File dialogue box

SETTING UP A DESKTOP THAT WORKS FOR YOU

When you set up a desktop, think carefully about how you will use your computer. Consider your needs as a writer—for instance, do you typically work on one project for a long time or do you work on several, often unrelated projects at once? Do you find yourself frequently using text that you've written before, such as standard parts of business letters or product descriptions? Consider too your strengths and weaknesses as a writer. Do you rely heavily on dictionaries or thesauruses while you write? Finally, consider issues such as whether you work alone or as part of a group, or whether you must follow particular style guidelines when you write.

After you've considered these issues, you'll be ready to design your desktop. Initially, you'll want to address two key questions:

- Will you need easy access to specific files and folders?
- Will you need easy access to specific software programs?

Accessing Files and Folders

Many writers find it easiest to place the folders and files they're working with directly on their desktops. Unfortunately, not all operating systems allow you to place files or folders on a desktop. Windows 3.x, for one, provides direct access to programs (through its Program Manager), but not to files. To access files and folders, you must either open them through a special program called File Manager or through your word processing program or other software programs.

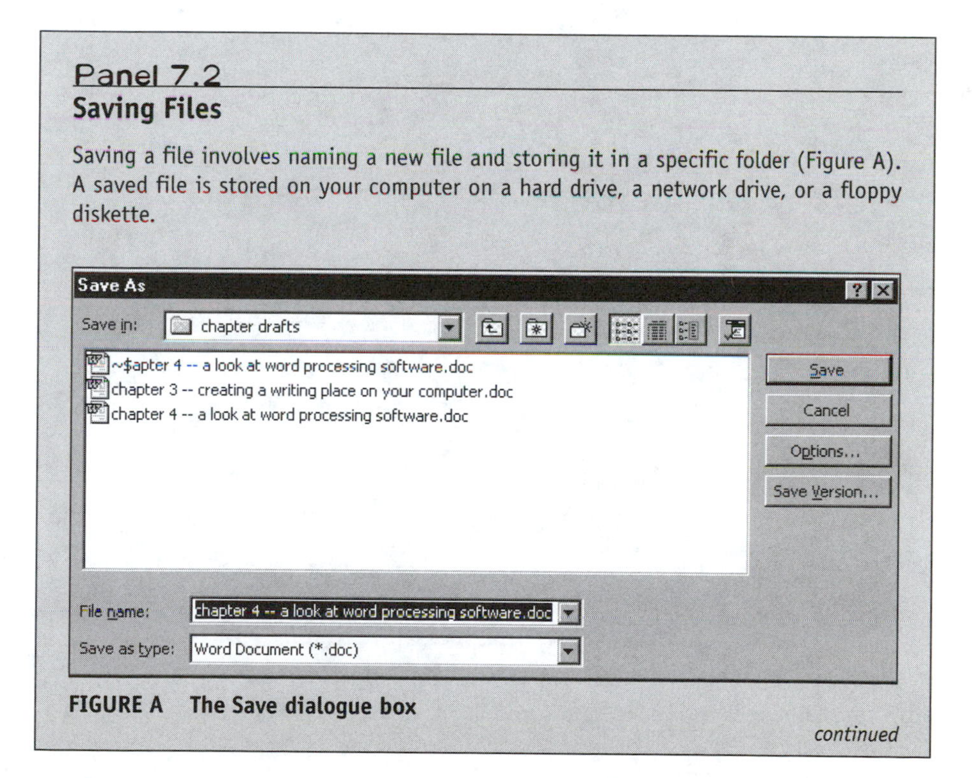

Panel 7.2

Saving Files

Saving a file involves naming a new file and storing it in a specific folder (Figure A). A saved file is stored on your computer on a hard drive, a network drive, or a floppy diskette.

FIGURE A The Save dialogue box

continued

Panel 7.2 *(continued)*

The most important rule to remember when working on a computer is "save early and save often." If you turn off your computer or quit your word processing program without saving a file, you'll lose some or all of your work. You'll also lose work if the power goes out unexpectedly or a small child accidentally turns off your computer. Most top-of-the-line word processing programs allow you to set periodic intervals for making an automatic backup copy of your file—that is, your computer will save the file in case of disaster. This only works, however, if you turn on the feature (Figure B). Even then, you'll likely lose a few minutes of work.

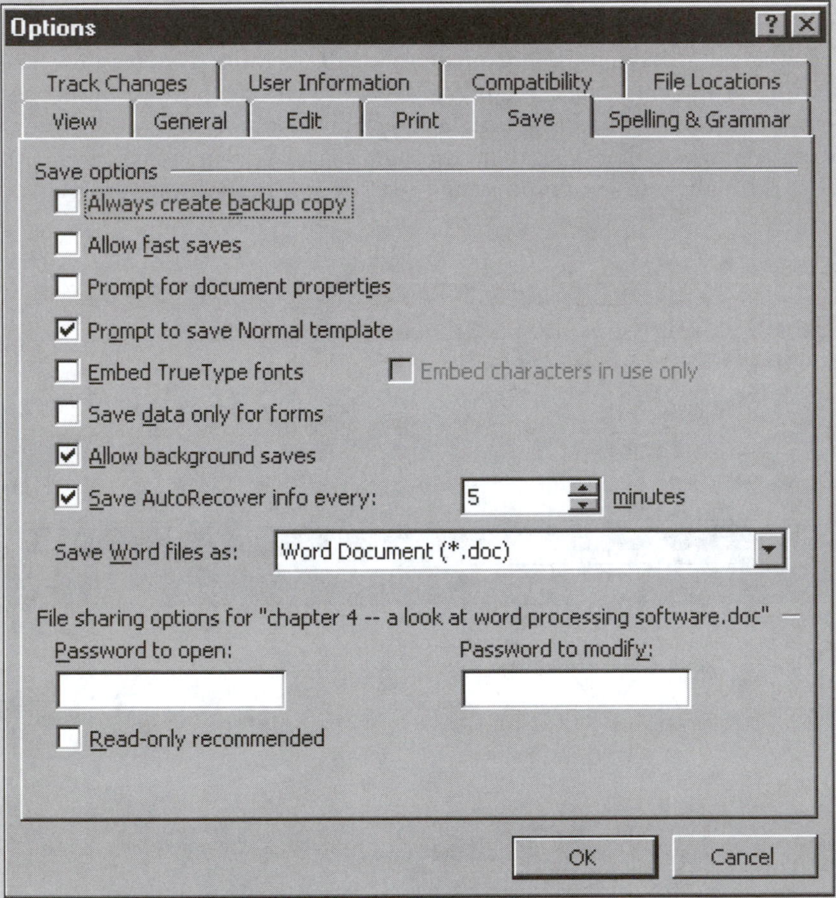

Options

Track Changes | User Information | Compatibility | File Locations
View | General | Edit | Print | Save | Spelling & Grammar

Save options
- [ ] Always create backup copy
- [ ] Allow fast saves
- [ ] Prompt for document properties
- [x] Prompt to save Normal template
- [ ] Embed TrueType fonts   [ ] Embed characters in use only
- [ ] Save data only for forms
- [x] Allow background saves
- [x] Save AutoRecover info every: 5 minutes

Save Word files as: Word Document (*.doc)

File sharing options for "chapter 4 -- a look at word processing software.doc"

Password to open: Password to modify:

- [ ] Read-only recommended

OK   Cancel

FIGURE B Turning on the AUTORECOVERY OPTION

Panel 7.2 *(continued)*

When you save a file, your word processing program will save it in a proprietary, or native format. In an ideal world, all word processing programs would save using the same format and one word processing program could open a file created in another without any difficulty. In this world, however, that doesn't happen. Most word processing programs can translate the format of another word processing program when you open or save a file (Figures C and D). However, this translation—like any translation between spoken languages—is seldom exact. Many writers find that translation among word processing formats is one of the biggest headaches they face when they share files with others. If you're working with others who have different word processing programs or even different versions of the same word processing program, try a few test translations to see what problems you might encounter when you try to share files.

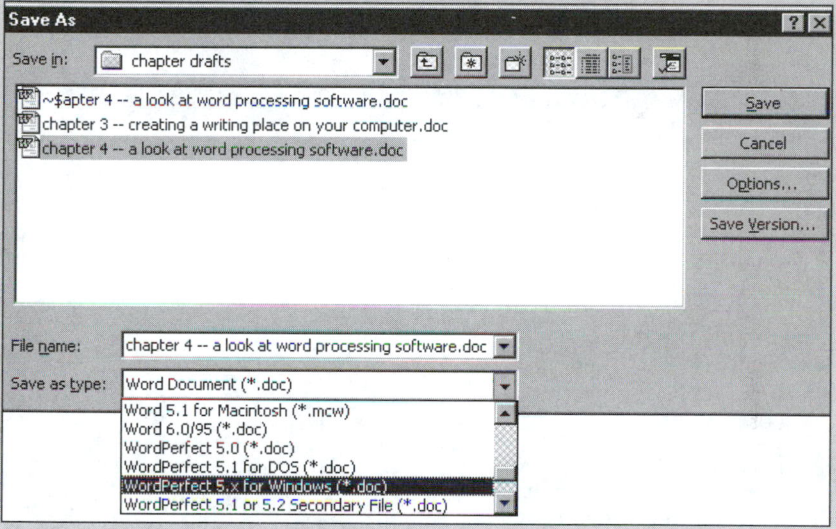

FIGURE C Specifying file format while saving a file

continued

If you're using Windows 95, Windows 98, Windows NT, OS/2, or the Macintosh Operating System, you can place files and folders on your desktop. Some writers reserve part of their desktop just for files and folders, while others let their folders and files fall where they may. Regardless of which decisions you make, you'll find that they will have an impact on your ability to easily locate and use files and folders.

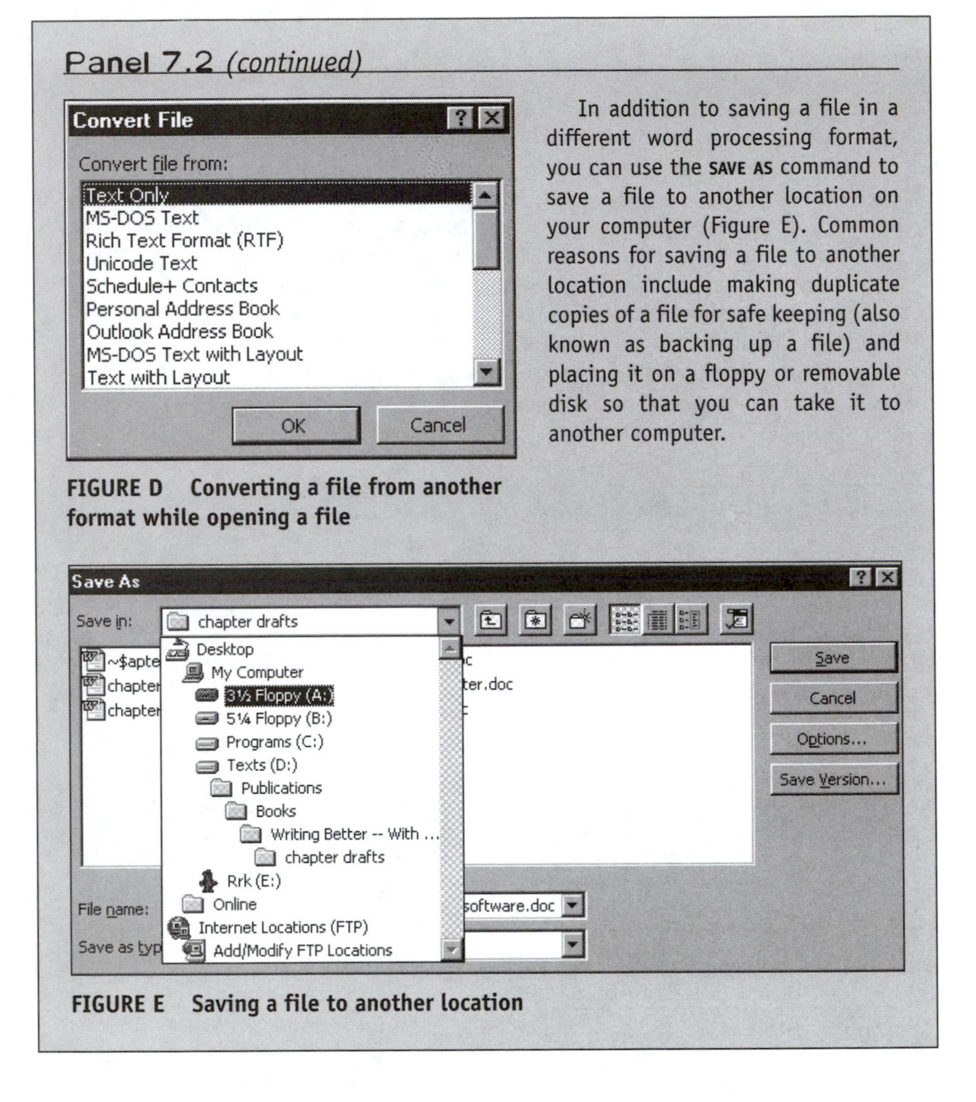

In addition to saving a file in a different word processing format, you can use the **SAVE AS** command to save a file to another location on your computer (Figure E). Common reasons for saving a file to another location include making duplicate copies of a file for safe keeping (also known as backing up a file) and placing it on a floppy or removable disk so that you can take it to another computer.

FIGURE D Converting a file from another format while opening a file

FIGURE E Saving a file to another location

Accessing Programs

The Macintosh Operating System marked a turning point in the way writers worked with files and folders. Until Windows 95, Windows 98, Windows NT, and OS/2, users of other personal computers used their software programs to work with files and folders. If you use Windows 3.x, for example, you probably start your word processor first and then open your files.

With the Macintosh OS, OS/2, and the newest versions of Windows, however, you can simply open a particular file. When you open the file, the appropriate program—your word processor, perhaps—automatically starts and opens the file. As a result, some writers place only file icons on their desktops. This approach works well if you only open files with particular programs.

If, however, you want to open the same file with different programs (which might happen if you want to view a file in one program, such as Netscape Navigator or Microsoft Internet Explorer, and edit it in another, such as Corel WordPerfect or a text editor), you'll want easy access to your program files. This can be done by placing them directly on your desktop or in folders on your desktop. In some operating systems, you can also assign programs to menus that can be accessed from the desktop (as is the case with the newest versions of Windows and the Macintosh OS).

Choosing Files and Programs to Place on Your Desktop

Choosing which files, folders, and programs you place on your desktop shapes how you work with your computer. So start simple and change your desktop over time as you polish your computer skills. You can get a good start, however, by asking the following questions:

- Do I want to directly access any of my files? Or do I want to access them through my software programs?
- If I want to directly access some of my files or folders, which ones should I put on the desktop?
- Do I want to access programs from my desktop? Or do I want to access them through other means, such as the Start button in Windows 95, Windows 98, or Windows NT?

If you want to put programs on your desktop, ask yourself:

- Which word processing software do I use the most?
- Do I want to have access to more than one kind of word processing software?
- Does the word processing program I use most frequently provide access to writing tools—spelling checkers, grammar and style checkers, thesauruses and dictionaries? If not, do I have access to these tools in the form of standalone programs?
- Do I need access to a Personal Information Manager (containing an address book, an appointment calendar, to-do list, and other functions)?
- Do I have a need for bibliographic programs (which support the insertion and formatting of literature citations)?
- Do I need access to text databases or programs that allow me to search for particular information in the files on my computer?

Panel 7.3
Naming Files and Folders

What's in a name? Plenty, if you want to figure out what's in one of your files or folders. Naming files and folders is one of the most important things to consider if you work with a large number of files and folders. At one time or another, most writers have tried without success to remember what they've named and where they've saved a particular file. It's a frustrating and annoying experience.

Many writers have also had the experience of trying to figure out which one of several similarly named files contains the document they're looking for. This is particularly a problem when working with the DOS or Windows 3.x operating systems which don't allow file names to have more than 11 letters or numbers. Short file names limit your ability to give descriptive names to your files. The result is often files with similar names, such as *mssoct10.doc, mssoct11.doc,* and *mssoct12.doc.* These files appear to contain manuscripts written in successive days in October, but it's not clear what they contain.

Devising a naming system that allows you to easily locate a file and determine what it contains will enhance your sense of inner peace and tranquility—or at least keep you from getting needlessly frustrated. We recommend the following principles:

- Use descriptive names within the limits of your operating system. Remember that file names in DOS or Windows 3.x are limited to 11 characters and that early Macintosh operating systems limit names to no more than 32 characters.
- Once you've created a naming system, use it consistently. Even a mediocre approach to naming files is better than three or four good, but different approaches when used together.
- If you work with large numbers of files, use a hierarchical approach to organize your folders. Group related files and folders in groups of increasing specificity. For instance, if you create a folder called *Letters,* you can create folders within *Letters* called *Letters of Recommendation, Letters to Clients,* and *Personal Letters.* Within *Letters to Clients,* you can create a folder for each client, or you can include the client's name as part of each file name. This approach to organizing your letter files should make it easy to find a particular letter sent to a particular client.
- If you are using DOS or Windows 3.x, use folders to indicate what is in a set of files. Using the previous example, your *Letters* folder would contain three folders named *Recs, Clients,* and *Personal.* Instead of putting all of your letters to clients in a single folder, create folders for each client and, within each of those folders, create additional folders corresponding either to the date the letters were sent or the content or purpose of groups of letters.

If you plan to share files among people using different operating systems, consider the file naming limitations of each system. Since DOS or Windows 3.x can't read long file names, the descriptive names on systems such as Macintosh, OS/2, or the other versions of Windows will be truncated. Before you begin sharing files, devise a naming scheme that will allow you to easily identify what's in the files.

- Which of the following communication programs do I regularly use?
    - Web browsers
    - Electronic mail programs
    - Electronic newsgroup readers
    - Electronic chat programs
- Do I regularly use any of the following tools?
    - File managers
    - File synchronization programs (for laptops and home or office PCs)
    - Utility programs, such as hard disk tools
    - Backup software
    - Graphics editors
    - CD players
    - Calculators

Your Computing Environment

Your computing environment also depends on how you organize files and folders on your computer, how you think about the relationships among files and folders, and the size and capabilities of your computer monitor.

Some people think about files and folders in a hierarchical sense—as a set of levels nesting within one another. Others think in terms of specific projects or activities. Still others don't differentiate among files and throw them all in a single folder, or perhaps a few folders. Experience suggests that the latter approach is a recipe for disaster. If you've ever scrolled endlessly through a list of files wondering what you named a file and where you put it, you know the problem.

In general, the most efficient way of organizing files is the hierarchical approach. This method involves creating a few broad categories at the top and gradually filtering down to the level of specific projects or topics (Figure 7.5).

Panel 7.4

Protecting Your Files and Folders

If you want to keep others from reading your files, you can protect them by assigning passwords. When a file is password protected, it can only be opened or copied by someone who knows the password. To find out how to set passwords on your files and folders, you should consult the help system that comes with your operating system or word processing software.

If you want to keep others from changing your files and folders, you can set their properties to read only. When files and folders are set to read only, they can be opened and read, but not changed.

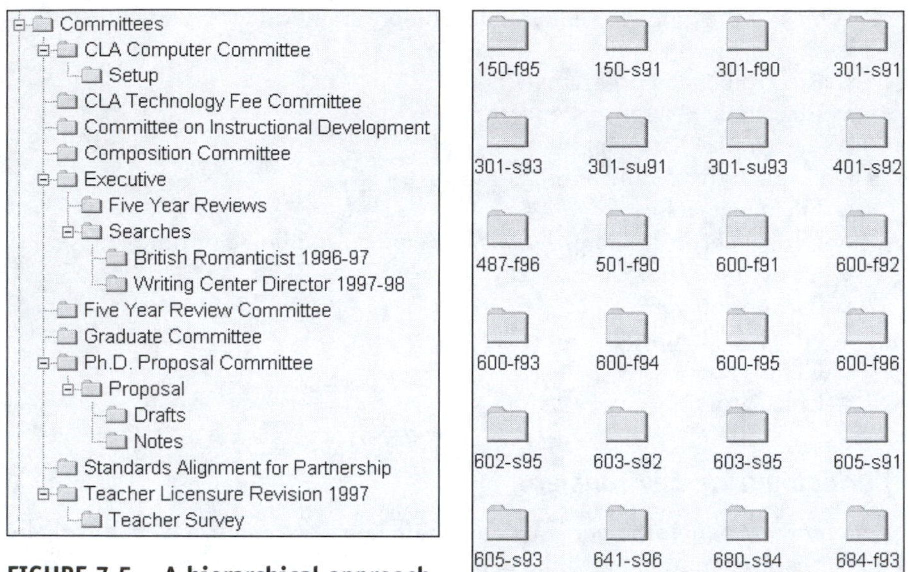

FIGURE 7.5 A hierarchical approach to organizing files and folders

FIGURE 7.6 A categorical approach to organizing files and folders—files from each course are saved in separate folders

The categorical approach works nearly as well, if you don't end up with too many categories (Figure 7.6). In no case, however, should you use a "flat" file system in which you put all of your files in a single folder. Although you'll no doubt know where all your files are, figuring out which files are which will take up an increasing amount of your time.

LOOKING AHEAD

Most writers spend little or no time creating a writing environment that supports the way they write. With a small amount of time and effort, however, you can create a desktop that increases your efficiency and productivity. If you take the time to do so, you'll find it much easier to locate the programs and files you need. And the less time you spend trying to remember where something is located on your computer, the more time you'll have for the real work of writing.

chapter 8

Creating a Writing Environment

When Sam moved to a new office, he placed his computer table and computer against the window. Finally, he could take a break from writing by looking out the window to watch the birds, squirrels, and early winter sunsets. His desk was to his right, so he could easily reach his references, telephone, and supplies. Within weeks, however, Sam's writing productivity had dropped. After writing for an hour, Sam felt tired, irritable, and uneasy.

Sam couldn't figure out what had happened. He needed to get his reports out, but they just weren't happening. Sam called in a consultant, Rosa Martinez. Rosa spotted some possible causes of Sam's sudden drop in writing productivity: the high contrast between the computer monitor and window background, a high-pitched 60 cycle hum from the air conditioning and heating unit, and overhead lights that cast glare on his computer monitor.

Your ability to write effectively and without unnecessary fatigue can be affected by a number of factors, including your surroundings, your physical makeup, and your personal habits and preferences. In this chapter, we explore the contributions these factors make to your writing environment—the place where you do most of your writing.

COMPUTER DANGERS—REAL OR IMAGINED

When Andre settled in before his computer, he felt fine, but within minutes Andre felt a nagging headache. After using her computer for a couple of hours, Emily's right hand felt numb. Juanita woke about midnight with her hand clinched tight in pain.

Millions of computers sit on desks in schools, business and industry, government agencies, and other organizations. Although it's clear that they've benefited students and workers in numerous ways, those benefits have not come without cost. A growing number of people attribute health problems to their heavy use of computers. Although research studies have not proven conclusively that computers cause problems, a prudent approach centers on understanding possible causes of the health problems among computer users and implementing preventive measures. The leading problems attributed to computers include:

- Repetitive stress injuries
- Fatigued muscles
- Eye strain and headaches

Repetitive Stress Injuries

Look around at meetings, conventions, and gatherings of computer users, and you'll see people wearing wrist braces. Talk to heavy computer users and you'll hear comments about sore wrists, elbows, arms, necks, and backs—usually from people suffering from repetitive stress injuries. If you perform the same task over and over again, you too may be at risk for repetitive stress injuries. Repetitive stress injuries include tendonitis—inflammation and swelling of tendons in hands, wrists, and arms; epicondylitis—inflammation and swelling of the tendons that join the elbow to the forearm muscles; de Quervain's disease—tendonitis of tendons connected to the thumb; and carpal tunnel syndrome—inflamed tendons that swell and compress the median nerve in the wrist (Furger 1993).

Repetitive stress injuries occur when you repeat the same motion over and over, often hundreds of times a day. The tendons rub together against each other and against sheaths or against bones, or they create pressure on nerves. To illustrate, take a quarter-inch wide rubber band, stretch it, and then pull it back and forth across the edge of a desk. The friction quickly heats the rubber band. Rub the band long enough and it will eventually break. In some ways, your tendons are like a rubber band. As you complete repetitive tasks, your tendons rub together and against your bones and muscle sheaths. As a result, they can become irritated and swell. The swelling puts pressure on nerves. Fortunately, unlike the rubber band, tendons do not break.

Typing for extended periods of time involves repetition of the same motions hundreds of times a day and may cause repetitive stress injuries in some people. Consider the common QWERTY keyboard—named after the arrangement of the top, left-hand row of keys. It was designed in the 1800s to slow typists down when using mechanical typewriters (Figure 8.1). According to studies completed by August Dvorak in the 1920s, the QWERTY keyboard forces typists to move their fingers between 12 and 20 miles during a typical 8-hour day (Cassingham, 1986). Dvorak studied the keystrokes and keystroke combinations required for

FIGURE 8.1 Arrangements of the QWERTY keyboard

typing in American English. He then rearranged the letters and produced a simplified keyboard (Figure 8.2). Type all day using a Dvorak keyboard and your fingers will only travel about one mile (Cassingham 1986).

Unfortunately, adoption of the simplified keyboard, often called the Dvorak keyboard, has not been widespread. Cassingham (1986) suggests three reasons. First, Dvorak introduced his simplified keyboard at the height of the Great Depression, and companies did not have sufficient capital to purchase new typewriters. Second, during World War II, typewriter plants were converted to small arms (pistol and rifle) production. Third, companies and government agencies saw the conversion to the Dvorak keyboard as too costly in terms of replacing the existing typewriters and retraining typists.

Despite the low adoption rate of the Dvorak keyboard, the American National Standards Institute (ANSI) approved the Dvorak keyboard layout in 1982, and published the standard as ANSI X4.22-1983. The advent of computers, ergonomic keyboards, and software programs now enables many computer users to convert to the Dvorak keyboard with a few clicks of a mouse. To learn more about the Dvorak keyboard, search the World Wide Web using the terms *Dvorak* and *keyboard.*

FIGURE 8.2 Dvorak, or simplified keyboard

tip 8.1 ■ Early Warning Signs

If you experience any of the following symptoms, don't delay. Seek medical help immediately:

- Numbness, tingling, or prickly feeling in your fingers, hands, wrists, arms, shoulders, neck, or back
- Muscle aches
- Backache
- Waking up in the middle of the night with shooting pain in your wrists, elbows, arm, or neck
- Headaches
- Squinting
- Eye focusing problems
- Scratchy eyes
- Watery eyes

In addition to his simplified keyboard, Dvorak also developed keyboards for one-handed typists. Thus, people with the use of only one hand can type efficiently. Microsoft provides information on Dvorak conversion software programs and one-hand keyboard programs that can be downloaded from its home page <http://www.microsoft.com>.

Not only does the arrangement of keys play a possible role in repetitive stress injuries, but also the horizontal keyboard height and vertical alignment of keys (looking at the keyboard from the top) may contribute to the problem. If your

Panel 8.1
Avoiding Repetitive Stress Injuries

Carpal tunnel syndrome and tendonitis are types of repetitive stress injuries. Symptoms include swelling, pain, and tingling in your hand or arm. Correct wrist position can prevent repetitive strain injuries.

Wrists should be neutral (straight). When you type, let your wrists float above the wrist rest. When you aren't typing, rest your wrist on the pad (see Figure A). Keep your wrists level. Avoid bending your wrists up or down while you type (see Figure B). While typing, don't rest your wrists on the edge of the keyboard or the table. This creates pressure points that can cause injury (see Figure C).

*Text used with permission. From "Preventing Computer Injuries" by Marla C. Roll from the Assistive Technologies Laboratory, Colorado State University, Fort Collins, Colorado, and with assistance from Pat Berlig, physical therapist, Colorado State University, Fort Collins, Colorado 80523.

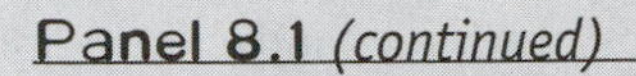

Panel 8.1 *(continued)*

FIGURE A Correct wrist position

FIGURE B Incorrect wrist position #1

FIGURE C Incorrect wrist position #2

keyboard is either too high or too low, you may flex your wrists up and down too much and the tendons in your wrists may rub against the bones and sheath more than they should. The standard straight-line alignment of the standard keyboard also causes you to flex your wrists right and left—an angle that may cause problems for some people.

In addition to concerns related to keyboards, some computer users have reported problems with their elbows and shoulders when using the mouse. Like problems associated with the keyboard, such problems may be associated with the position your hands, wrist, and arms are in while you use and reach for the mouse.

Fatigued Muscles

If you write under pressure—for example, you've put off writing until the last minute and you write two, four, or more hours without breaks—you'll probably experience muscle fatigue. When people work under pressure, they tighten their muscles and muscle fatigue soon sets in. Aching muscles often result.

If you take a dictionary and hold it at arm's length as long as you can, you'll soon fatigue. You've had to tense your muscles and hold them in the extended position without rest. But you could hold the dictionary at an arm's length for a much longer time if you could take frequent breaks. Similarly, you can avoid muscle fatigue by avoiding marathon writing sessions. As you write, vary your activities at the computer, take frequent breaks, and engage in stretching exercises.

If possible, avoid marathon writing sessions by varying your activities at your computer desk. Because more and more writing and related information gathering and management activities can be done on the computer, you may find yourself spending two, four, six, or more hours working at the computer. Prior to computers, writers moved about. They varied their activities. Some wrote longhand, and then typed their drafts. Others, such as journalists, composed directly at the typewriter. Once they had a typed article, they would turn to their desks, pick up a pencil and begin copyediting their article. They'd pull a dictionary from the shelf, look up words, and enter correct spellings as they copyedited their article. Finally, they would then retype the article to reflect their copyediting changes.

When working at a computer, plan to break up your writing sessions with other activities. Make telephone calls, read sections of books and articles, or take a coffee break. By varying your activities, you won't keep the same muscles in the same position and you won't use the same muscles over and over again. Simply put, you'll relax the muscles you're using when writing.

You can relax your muscles by taking breaks and stretching. See the reprint of Bob and Jean Anderson's *Computer & Desk Stretches* at the close of this chapter. They've prepared *Computer & Desk Stretches* in an 8.5- by 11-inch laminated card so computer users can keep it near their desk to guide their stretching. And they have also written a brief book, *Stretching at Your Computer Desk*, and a companion software program. For more information, contact Stretching, Inc.,

P.O. Box 767, Palmer Lake, Colorado 80133-0767, telephone 1-800-333-1307, or <http://www.stretching.com> on the World Wide Web.

Headaches and Eye Strain

Reading text on screen is more difficult than reading printed copy. In part, it comes from the nature of the computer monitors and how monitors are typically positioned.

The Nature of Computer Monitors. Reading from a computer monitor is unlike reading printed page for several reasons: image quality, viewing distance, projected text, screen resolution, and flicker.

The print on a computer monitor does not approach the quality of text printed on paper. To illustrate, take a printed page, and hold it next to your computer screen. Look closely at the text on the computer screen. Note the differences in text quality. With many monitors, the monitor's curved surface distorts, sometimes slightly and other times grossly, the text on the far right and far left edges of the screen.

Reading from a computer monitor makes your eyes work harder. Whether reading printed copy or hard copy, your eyes must focus on objects up close. To do so, your eye muscles contract and reshape your eye allowing you to refocus for the closer distance. If you write on a computer for hours, the need to maintain the close focus stresses the eye muscles in much the same way that reading printed copy for hours stresses your eyes.

When you read from the computer screen, you're also reading projected text rather than reflected text. On some monitors, the projected text creates a halo or shadow-like appearance around letters. The letters may appear fuzzy. Because most of your reading has been on printed copy that is sharper, your eyes may struggle to refocus the text. If you read or write for long periods of time without breaks, you may find it stressful.

Text size also plays an important role in the ease of reading. In monitors, size matters more than any other factor in determining the best text size and resolution or sharpness. Although you will gain more screen "real estate" by running at a higher resolution, you'll find that your fonts will appear smaller. You may find the smaller text stains your eyes. If, like many writers, you find that you have to spend long stretches of time staring at your monitor, you may find it best to use a slightly lower resolution (say 800 by 600, rather than 1,024 by 768). By using a lower resolution, the text will be larger on your monitor. If you use a higher resolution, but are worried about eye strain, try using larger than normal screen fonts, such as 12- or 14-point type. Although the programs will still display at the higher resolution, larger fonts may allow you to read on-screen text more easily.

When considering monitors, keep in mind their refresh rate. Conventional computer monitors use electron guns to "refresh the screen" many times each second. Most people perceive a lower refresh rate (usually below 70 times per second) as flicker. The lower the refresh rate, the higher the perceived flicker. Unfortunately, many inexpensive computer monitors exhibit a discernible flicker when set to display a higher resolution (such as 1,024 by 768 or higher). Not only might you find such flicker annoying, it may lead to eye strain and headaches. If you notice that your monitor is flickering, reduce the display resolution. To do this, check your operating system's online help, its manual, or your video adapter manual for guidelines on how to reset the resolution on your computer.

A multitude of monitor problems can contribute to eye-strain and headaches. If you experience these symptoms, try adjusting your monitor and software setting to minimize the problems.

Positioning Your Monitor. When you look at a computer screen, your eye muscles contract so that you can focus on closer objects. When you glance away from the computer screen, your eyes relax, as long as you focus on a distant object. But far too many computers are located against walls. In such cases, glancing away from the computer does not allow your eyes to relax as much as they would if the monitor had space behind it.

Unfortunately, it's usually not a good idea to position your monitor against a window so you can look outside. For many computer users, placing a computer next to a window can cause problems with contrast, the difference in light intensity between the computer background and the area behind the computer. Similarly, overhead lights, doors, and other light sources can contribute to contrast problems. In addition, light sources in a room can create problems related to glare, and the reflection of light off the computer screen into your eyes. Sources of glare include light from windows, overhead lights, open doors, other rooms, and glass dividers.

When your eyes encounter glare, your pupils contract, and you're likely to squint. Consider the situation where you're driving into the sun in the early morning or late afternoon. When glare strikes your windshield, you squint in an attempt to see through the windshield. In most cases it lasts for only a few seconds. Although the glare off of most computer screens is seldom as intense as glare from the sun, computer glare may last for hours. Although you may not realize you're squinting, you can easily develop headaches. Only when you position your computer properly or adjust room lights can you eliminate glare.

ARRANGING AND ADJUSTING YOUR WRITING ENVIRONMENT

Prevention remains the key to reducing possible risks from computers. By understanding the possible risks, planning ahead, and taking corrective actions if symptoms arise, you can minimize the chances of long-term injuries. To reduce the chances of possible repetitive stress injuries, muscle aches, eyestrain and headaches, find ways to optimize your writing environment. Some of the most important steps you can take include obtaining an adjustable chair, obtaining an adjustable computer desk, arranging your computer table and desk, adjusting your chair and computer desk, adjusting your keyboard, adjusting the monitor height and distance, and adjusting room lighting.

Obtaining an Adjustable Chair

Optimizing your writing environment begins with an office chair that fits your body. Begin with an office chair. It should have, at a minimum:

- Five legs for stability
- Adjustable seat height
- Adjustable arms
- Adjustable back and lumbar support

Chairs with five legs are highly stable; you'll have much less chance of tipping over and your falling out of the chair. Don't laugh. Chairs tip over with more frequency than you might imagine, and some people are seriously injured. Five-legged office chairs provide more stability than the older four-legged office chairs.

Your computer chair should allow you to adjust the height of the seat. You should be able to lower the seat so your legs form a 90 degree angle and your thighs and buttocks are parallel to the floor. When you sit in the chair, you should be able to put your hand between the front of the chair and the back of your knee. The front edge of the chair should drop away, much like a water-fall, so that it does not put any pressure on your thighs, the back of your knee, or the back of your calves.

If you happen to be short, some chairs cannot be adjusted low enough so that your feet touch the floor. If you adjust a chair to its lowest position and your feet still don't touch the floor, then you'll need a foot rest. You can buy them at computer stores, office supply stores, or mail order houses. Use a footrest to ensure that your legs are not dangling over the edge of the chair and putting pressure on your legs. In a pinch, use a small box until you can purchase a foot rest.

Some office chairs have arms, some do not, and some have adjustable arms. Sometimes chair arms interfere with your ability to properly position your arms when you're keyboarding. Your arms need to be free so that your elbows and forearms aren't hitting the chair arms as you type. Some writers prefer chairs without arms while others prefer chairs with arms. It's an individual call that depends on how well your body fits your chair, your computer desk, and your keyboard arrangement.

Obtaining an Adjustable Computer Desk

An ideal computer desk will allow you to adjust both the keyboard and monitor height to fit your body. Look for computer desks that allow you to adjust the keyboard and monitor height. Some desks come with incremental height adjustments about every inch while others allow continuous adjustments. Computer desks with continuous adjustments for the desktops have a crank and chain. As you turn the crank, it moves a chain that moves the computer table up or down. Such furniture works especially well if several people use the same computer desk. It also works well for people with repetitive stress injuries who need to fine-tune their keyboard and monitor heights to minimize further health problems.

As an alternative to an adjustable desk, use an adjustable keyboard and mouse arm that attaches under the computer desk top and swings up and out with incremental adjustments. Such arms need adequate space for your mouse so that you can move it around without overly extending your arm.

tip 8.2 ■ Reducing Noise in Your Writing Environment

If you prefer a quiet place to write, but must work in a noisy setting, explore strategies for cutting out the noise. Try the following strategies:

- Play music through headphones to screen out distractions
- Wear ear plugs
- Schedule your writing sessions early or late in the day
- Take a laptop computer to a quiet place, such as an empty conference room or a library

It is important that you learn to shut out the noise by concentrating deeply on your writing.

Considering Ergonomic Keyboards

Manufacturers have developed and experimented with a wide range of ergonomic keyboards designed to reduce the possibility of repetitive stress injuries. Most ergonomic keyboards are variations on a split keyboard set at different angles and positions to reduce the possibility of repetitive stress injuries.

Microsoft's Natural keyboard, an ergonomically designed keyboard, comes with a slip key arrangement, wrist leveler, software for converting the keyboard arrangement to the Dvorak keyboard, and a succinct manual with detailed ergonomic guidelines. The keyboard is moderately priced and available from many computer stores and mail order vendors.

Arranging Your Computer Desk and Office Desk

In an ideal arrangement, you need space to arrange your computer, reference materials, notes, telephone, and other resources so they're handy. In its guide accompanying the Natural Keyboard, Microsoft suggests three workzones: (1) a primary task zone, (2) a secondary work zone, and (3) a reference zone. Place your most frequently used items within the primary zone—the distance from your elbow to your hand. Place items that you use somewhat less frequently within the secondary zone, an arm's reach. Place your least used items in the reference zone, outside your arm's reach (Microsoft, 1994).

Some writers have found "L" and "U" shaped arrangements of their computer desk and office desk helpful (Figure 8.3). These arrangements provide plenty of space to arrange your computer, printer, telephone, reference materials, and resources. Some writers have even had desks built for their specific needs. The writer William F. Buckley, Jr., for instance, had a horseshoe shaped desk built for his home office (Robinson, 1994).

Under ideal conditions, place the computer desk so that your monitor will sit 90 degrees distance from any window and between any overhead lights. Don't place your computer against a window. The contrast between the monitor and background may present problems for you. Consider too the overhead lighting in the room. Try to arrange your desk so that you do not have overhead lights directly above you and your computer. Overhead lights sometimes create glare.

Adjusting Your Chair and Computer Desk

To fit your body to your workstation, first adjust your chair so that your legs touch the floor and form a 90 degree angle. If you're short, use a foot rest—make sure that your feet are not dangling in the air and creating pressure on your legs. As you sit, do not have pressure from the chair on the back of your knees, your thighs, or your calves. Try to slide your hand between the back of your knee and the chair; it should slide in easily. If not, readjust the chair. If possible, have a friend or colleague look at your chair from the side to make sure you have adjusted the chair to the correct height.

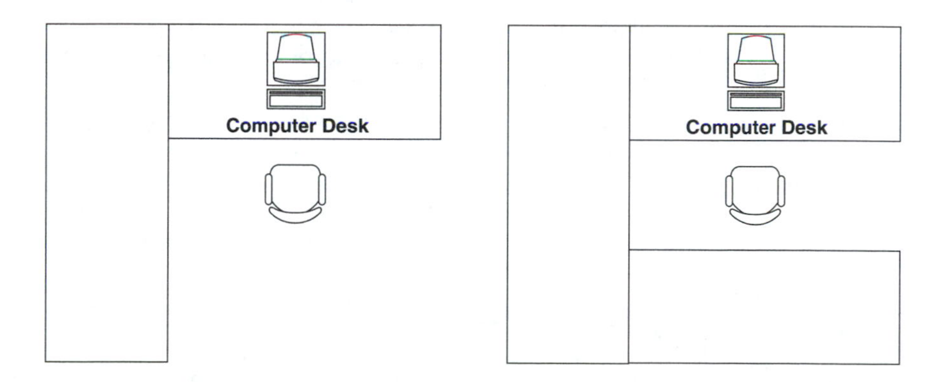

FIGURE 8.3 Suggested desk arrangement, viewed from above

Panel 8.2
A Recommended Working Posture

Screen: Slightly below eye level

Keyboard: Positioned to allow straight hand and forearm

Document holder: Same height and distance as screen

Viewing distance: 18–28 inches

Gaze angle: 35 degrees

Elbow angle: 90 degrees

High angle: 90 degrees

Wrist angle: Neutral

Normal head tilt: 15 degrees

Angle of screen: 88–105 degrees

Backrest height: 18–20 inches

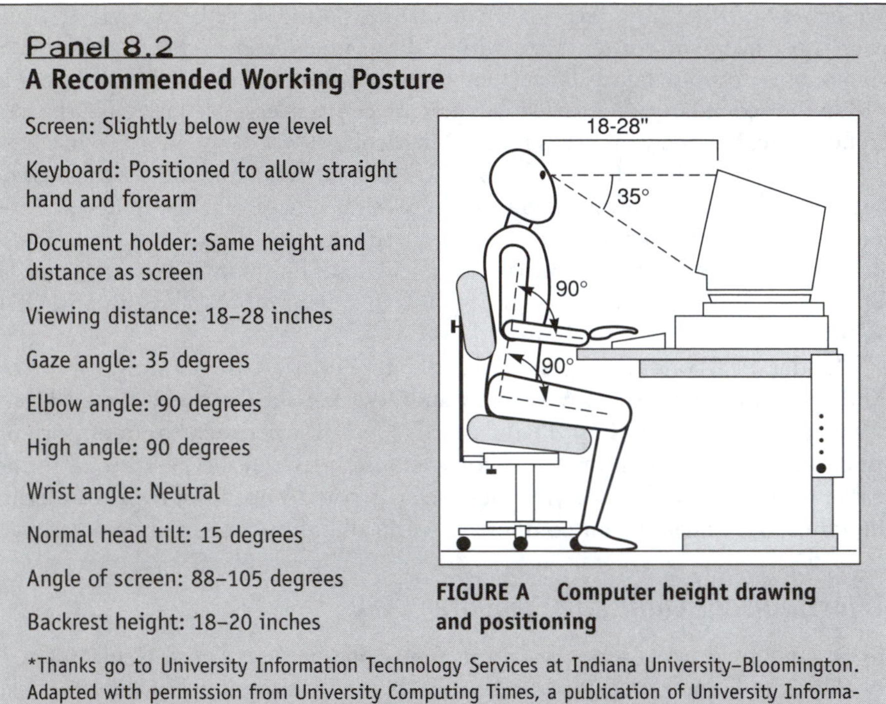

FIGURE A Computer height drawing and positioning

*Thanks go to University Information Technology Services at Indiana University–Bloomington. Adapted with permission from University Computing Times, a publication of University Information Technology Services at Indiana University–Bloomington. Also adapted with permission of Mark Sheehan, University of Montana.

Adjust the chair back so that it gives you solid lumbar support and presses gently against your back as you sit. Adjust the chair back so it's slightly inclined toward the rear to take pressure off the discs in your spine. Sitting puts more pressure on your spine and discs than most other activities, so careful adjustment can help reduce the pressure and back aches.

To adjust the height of the computer desk, let your arms hang loosely at your sides and raise your forearms so that they're parallel with the floor and your wrists are extended straight. Your hands should be nearly parallel to and slightly above the keyboard. Adjust the desk height so the keyboard slips easily below your hands. Your wrists should not flex up or down more than a few degrees.

Adjusting Your Keyboard and Mouse

With your arms tucked against your sides and your forearms extended, center your keyboard until you find a comfortable position just below your hands as your arms hang loosely at your side and your forearms and wrists remain parallel to the floor. Position the mouse as close to the keyboard as possible to avoid stretching to reach it.

Many keyboards have adjustable feet or levelers to help align the keyboard so you can keep your hands parallel and just above the keyboard. You'll need to try typing with the different keyboard adjustments to see if you can keep your wrists aligned with minimal up or down flex. Curl your fingers so you can gently touch the keys while your wrists remain straight as you type.

Adjusting Your Monitor

Place your monitor an arm's length from you—between 18 and 28 inches—and adjust its height so that you can look straight on or slightly down. If you wear bifocals, you'll need to lower the monitor so that you're not looking through the bottom lens—the focusing distance for the bottom lens ranges from 12 to 18 inches, depending upon your prescription.

If your desk has an adjustable shelf for a monitor, adjust the shelf so that your monitor will sit at the appropriate height. Place the monitor on the shelf, then sit down and try it out. Have a colleague stand to the side to see if it's at the correct height. If your computer desk does not have a shelf, you'll need to purchase a monitor stand, make one, or improvise. Many computer suppliers sell monitor stands. You can also build one from three-quarter inch plywood. Some writers improvise temporary monitor stands—they place telephone books and other large books under monitors. Although such approaches work, monitors can slip off make-shift stands and crash to the floor. The moral? Don't scrimp and save a few dollars on a monitor stand only to be billed for an expensive monitor repair or replacement.

If you're reading text and then writing a document, position a document holder at the same height as your computer monitor. Avoid having to twist your head, neck, and shoulders excessively as you write. Try to keep looking at the monitor or slightly to the side if you're reading notes, extracting information from other documents, or entering changes from a copyedited document.

Adjusting Your Lighting

Check your monitor for glare from windows, overhead lights, hallways, and other sources. You may need to turn off overhead and room lights or add anti-glare screens to your monitor. Some writers turn off the overhead lights and add desktop lights or area lighting to illuminate their desktops.

Glare from overhead lights is not a new phenomenon. Decades ago, newspaper editors wore green eye shades or visors to keep the glare from overhead lights out of their eyes as they copyedited stories and proofread galleys and page proofs. Although computer supply stores don't stock green eye shades today, some writers have resorted to wearing baseball caps to shade their eyes when they can't turn off the overhead lights. Other writers have taped narrow pieces of cardboard to the top of their monitors to cast a shadow on the screen and eliminate glare. They do so carefully, however. If they block the vents on the monitor, heat can build up and damage the monitor.

CHANGING YOUR WORK HABITS

Once you've arranged your writing environment, you'll want to develop good posture and work habits. In particular, you'll want to focus on sitting properly in your chair and keyboard, looking beyond your computer monitor frequently, taking frequent breaks, and avoiding marathon writing sessions.

Sitting Properly

Developing good posture at your computer desk takes practice. Some writers try to relax by slouching down, crossing their legs, putting their legs up on their chair legs, folding their legs under their chairs, and attempting various other contortions. Although it takes practice, sitting properly at your computer desk minimizes the chances of incurring health problems.

Try to keep your feet flat on the floor, sit back in your chair, and hold your arms so that your hands rest just above the keyboard. Type lightly, don't pound the keys. It takes little effort to make your key strokes appear on the screen. Don't rest your wrists or hands on any sharp edges on the desk.

If you use a wrist pad, use it to rest your hands between writing. Don't place the palms of your hands on the wrist pads and type. For many people, resting their wrists on the desk and typing creates an unhealthy angle that can lead to tendon damage.

Looking beyond Your Monitor

Take mini-breaks as you write and look beyond your monitor. If you have distance behind your monitor, focus on an object on the wall—a picture, clock, or other object. This helps relax your eyes and reduce the tension in your eye muscles. If your monitor sits against the wall, look to the side and across the room to relax your eyes. If you don't have anything on the walls, hang a picture, photograph or drawing on a far wall.

Taking Frequent Breaks

Every half hour, take a five minute break from your writing. Do something else, such as reviewing your notes, looking up other information, or editing a printed copy of your document. Do anything to give yourself a break from writing on the

Panel 8.3

Bob and Jean Anderson's Computer and Desk Stretches

Sitting at a desk or computer terminal can cause muscular tension and pain. Take a few minutes to do a series of stretches and your whole body will feel better. It is helpful to stretch spontaneously throughout the day, stretching any area of the body that feels tense. This will help greatly in reducing and controlling unwanted tension and pain. *(Most of these stretches may be done standing or sitting. When standing remember to keep your knees slightly bent to protect your back and to give you better balance.)*

How to Stretch:

- Stretch to a point where you feel a mild tension and relax as you hold the stretch.
- The feeling of stretch tells you whether you are stretching correctly or not.
- If you are stretching correctly, the feeling of stretch should slightly subside as you hold the stretch.
- Do not bounce.
- The long-sustained, mild stretch reduces unwanted muscle tension and tightness.
- Stretches should be held generally for 5–30 seconds, depending on which stretch you are doing.
- Breathe slowly, rhythmically and under control.
- Relax your mind and body as much as possible.
- Always stretch within your comfortable limits, never to the point of pain.
- Do not compare yourself to others. We are all different. Comparisons only lead to overstretching.
- Any stretch feeling that grows in intensity or becomes painful as you hold the stretch is an overstretch.

**Note: If you have had any recent surgery, muscle, or joint problem, please consult your personal health care professional before starting a stretching or exercise program.*

The dotted areas are those areas where you will most likely feel the stretch.

1

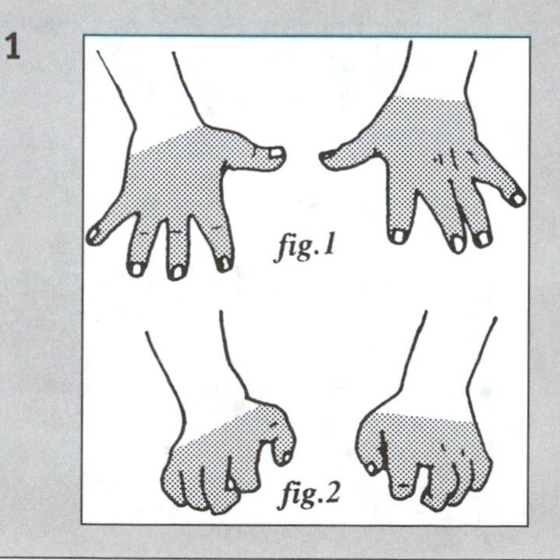

Separate and straighten your fingers until tension of a stretch is felt (fig. 1). Hold for 10 seconds. Relax, then bend your fingers at the knuckles and hold for 10 seconds (fig. 2). Repeat stretch in fig. 1 once more.

continued

Panel 8.3 *(continued)*

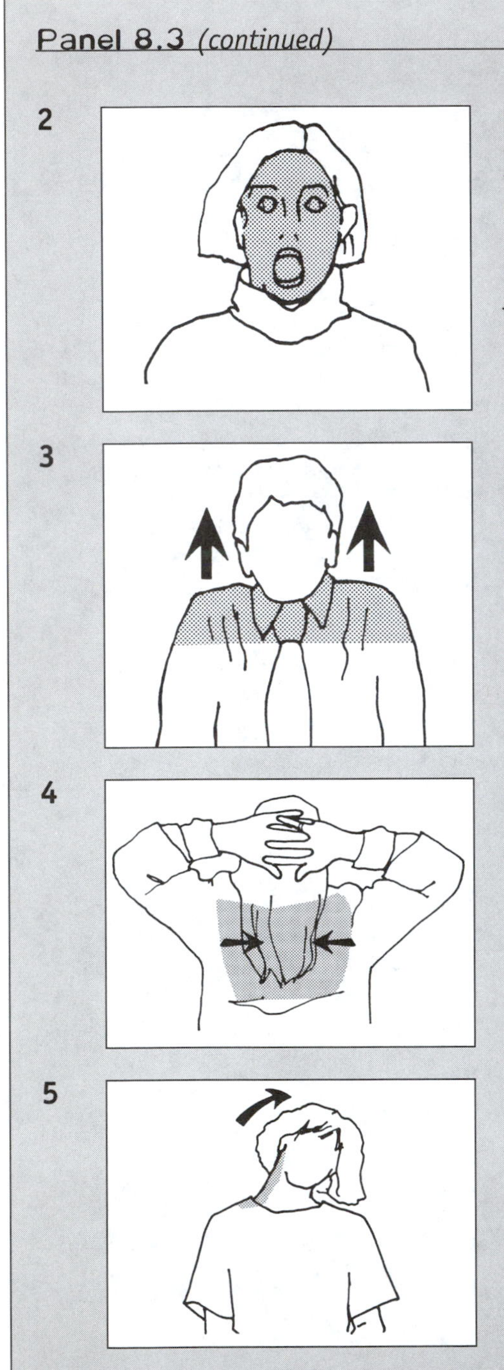

2 Raise your eyebrows and open your eyes as wide as possible. At the same time, open your mouth to stretch the muscles around your nose and chin and stick your tongue out. Hold this stretch for 5–10 seconds. *Caution: If you hear clicking or popping noises when opening mouth, check with your dentist before doing this stretch.*

3 Shoulder Shrug: Raise the top of your shoulders toward your ears until you feel slight tension in your neck and shoulders. Hold this feeling of tension for 3–5 seconds, then relax your shoulders downward. Do this 2–3 times. Good to use at the first signs of tightness or tension in the shoulder and neck area.

4 With fingers interlaced behind head, keep elbows straight out to side with upper body in a good aligned position. Pull your shoulder blades toward each other to create a feeling of tension through upper back and shoulder blades. Hold this feeling of mild tension for 8–10 seconds, then relax. Do several times.

5 Start with head in a comfortable, aligned position. Slowly tilt head to left side to stretch muscles on the right side of neck. Hold stretch 5–10 seconds. Feel a good, even stretch. Do not overstretch. Then tilt head to right side and stretch. Do 2–3 times to each side.

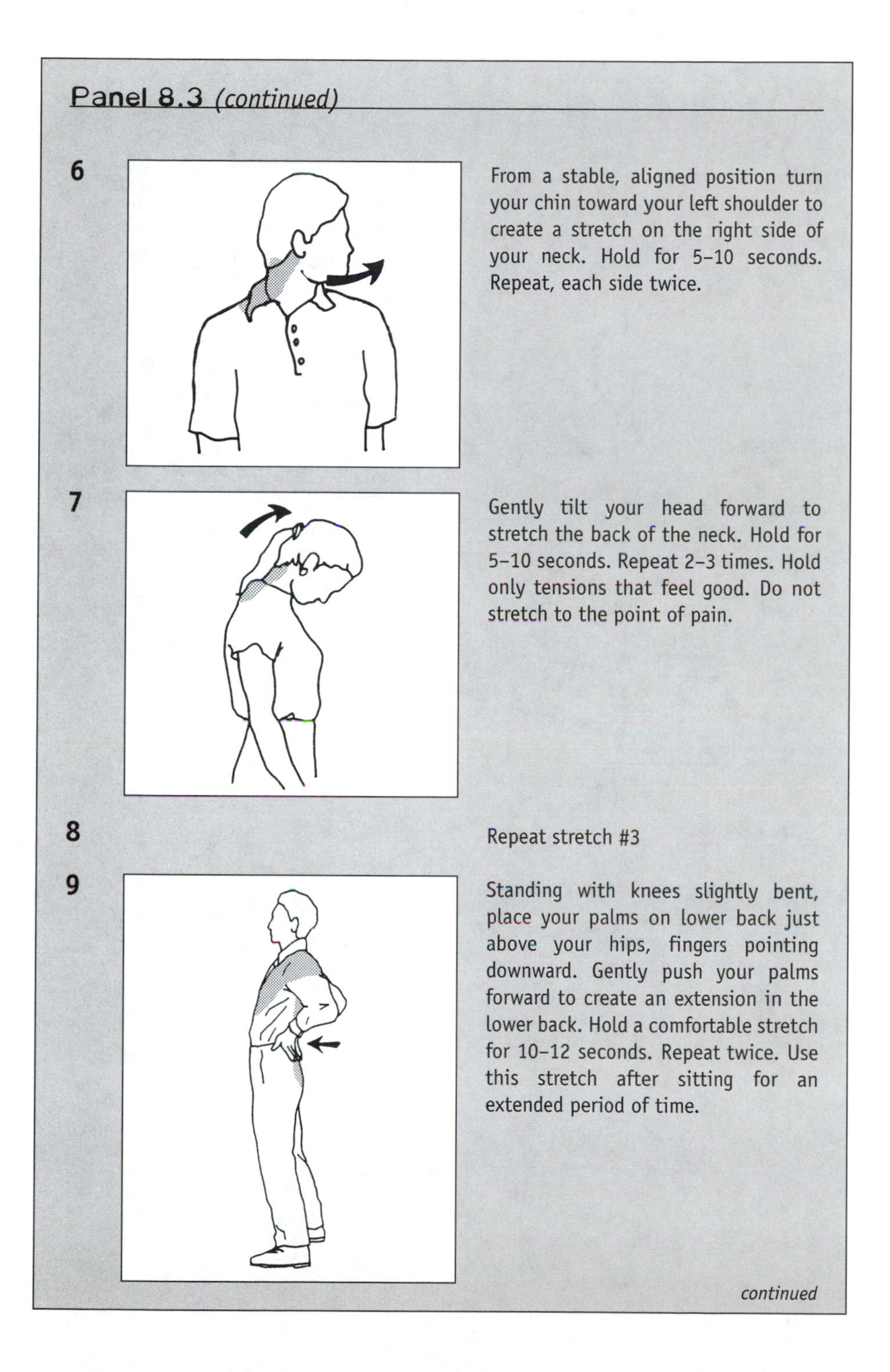

Panel 8.3 *(continued)*

6

From a stable, aligned position turn your chin toward your left shoulder to create a stretch on the right side of your neck. Hold for 5–10 seconds. Repeat, each side twice.

7

Gently tilt your head forward to stretch the back of the neck. Hold for 5–10 seconds. Repeat 2–3 times. Hold only tensions that feel good. Do not stretch to the point of pain.

8

Repeat stretch #3

9

Standing with knees slightly bent, place your palms on lower back just above your hips, fingers pointing downward. Gently push your palms forward to create an extension in the lower back. Hold a comfortable stretch for 10–12 seconds. Repeat twice. Use this stretch after sitting for an extended period of time.

continued

Panel 8.3 *(continued)*

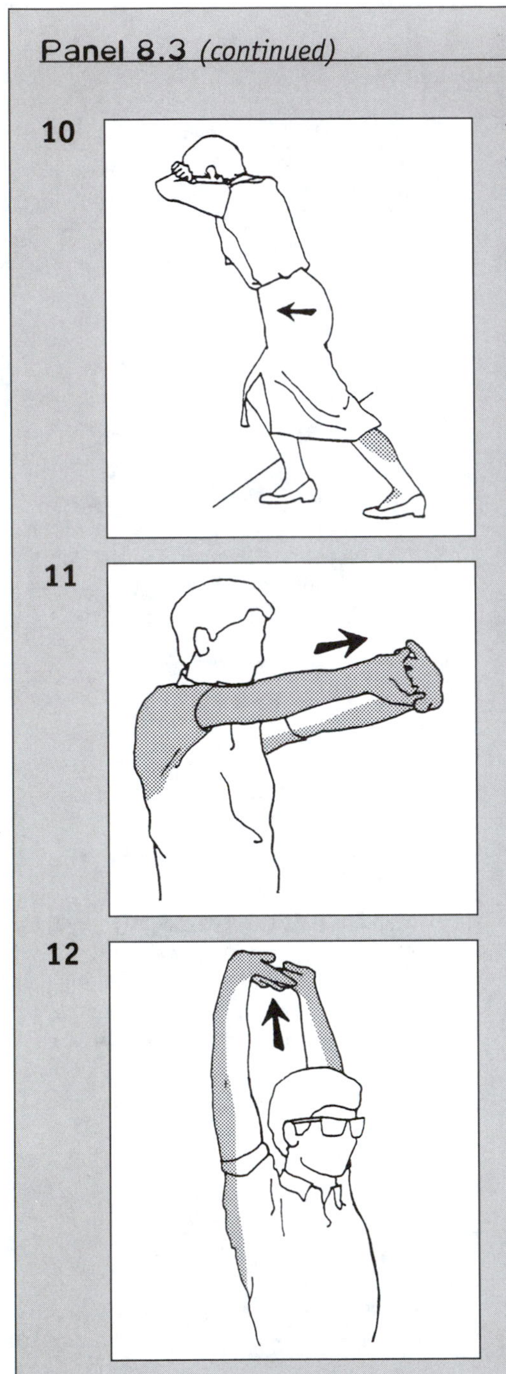

10 To stretch your calf, stand a little ways from a solid support and lean on it with your forearms, your head resting on your hands. Bend one leg and place your foot on the floor in front of you leaving the other leg straight, behind you. Slowly move your hips forward until you feel a stretch in the calf of your straight leg. Be sure to keep the heel of the foot of the straight leg on the floor and your toes pointed straight ahead. Hold an easy stretch for 10–30 seconds. Do not bounce. Stretch both legs.

11 Interlace fingers, then straighten arms out in front of you, palms facing away from you. Hold stretch for 10–20 seconds. Do at least two times.

12 Interlace fingers then turn palms upwards above your head as you straighten your arms. Think of elongating your arms as you feel a stretch through arms and upper sides of rib cage. Hold for 10–20 seconds. Do three times.

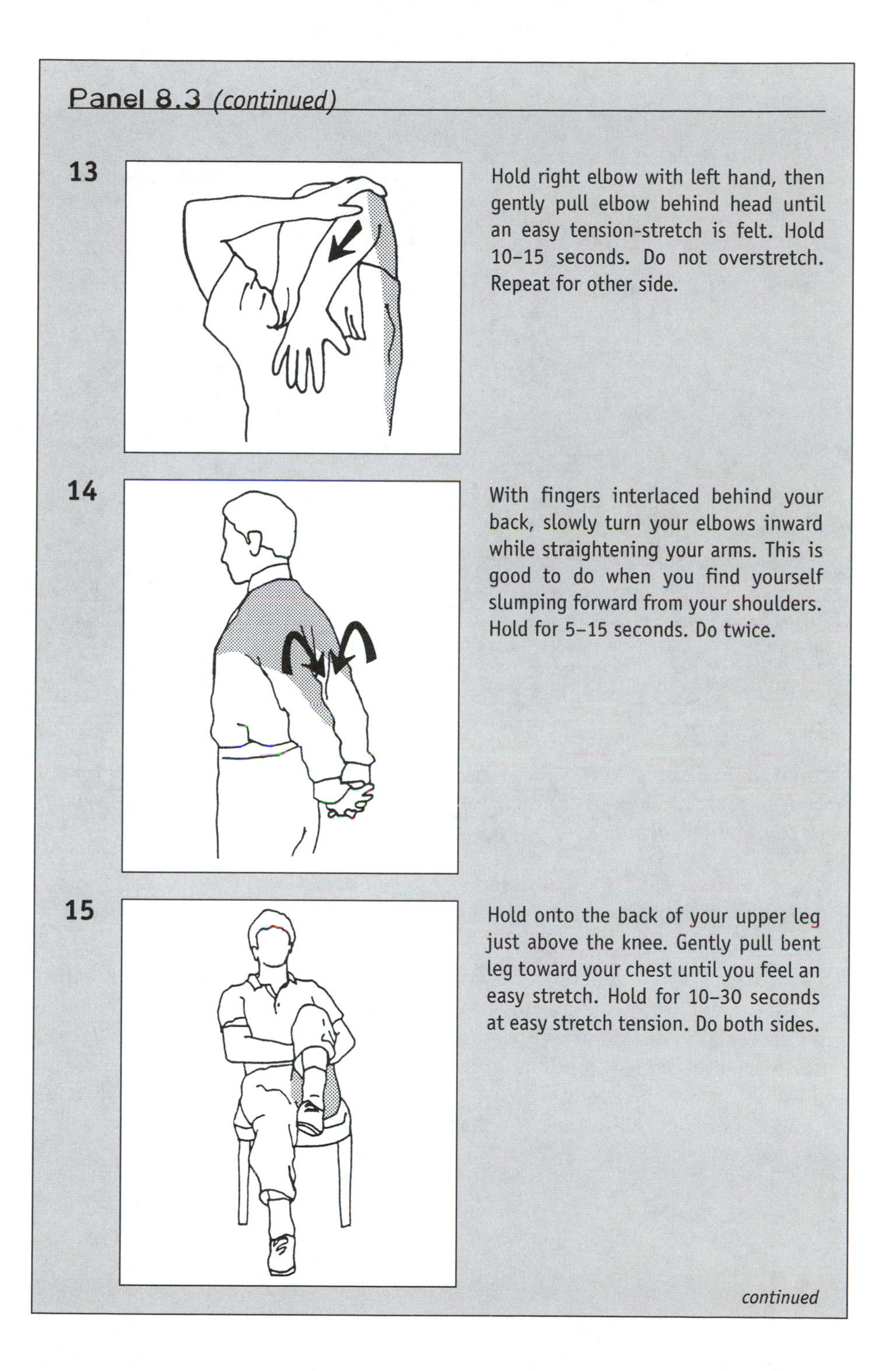
Panel 8.3 (continued)
13
Hold right elbow with left hand, then gently pull elbow behind head until an easy tension-stretch is felt. Hold 10–15 seconds. Do not overstretch. Repeat for other side.
14
With fingers interlaced behind your back, slowly turn your elbows inward while straightening your arms. This is good to do when you find yourself slumping forward from your shoulders. Hold for 5–15 seconds. Do twice.
15
Hold onto the back of your upper leg just above the knee. Gently pull bent leg toward your chest until you feel an easy stretch. Hold for 10–30 seconds at easy stretch tension. Do both sides.
continued

Panel 8.3 *(continued)*

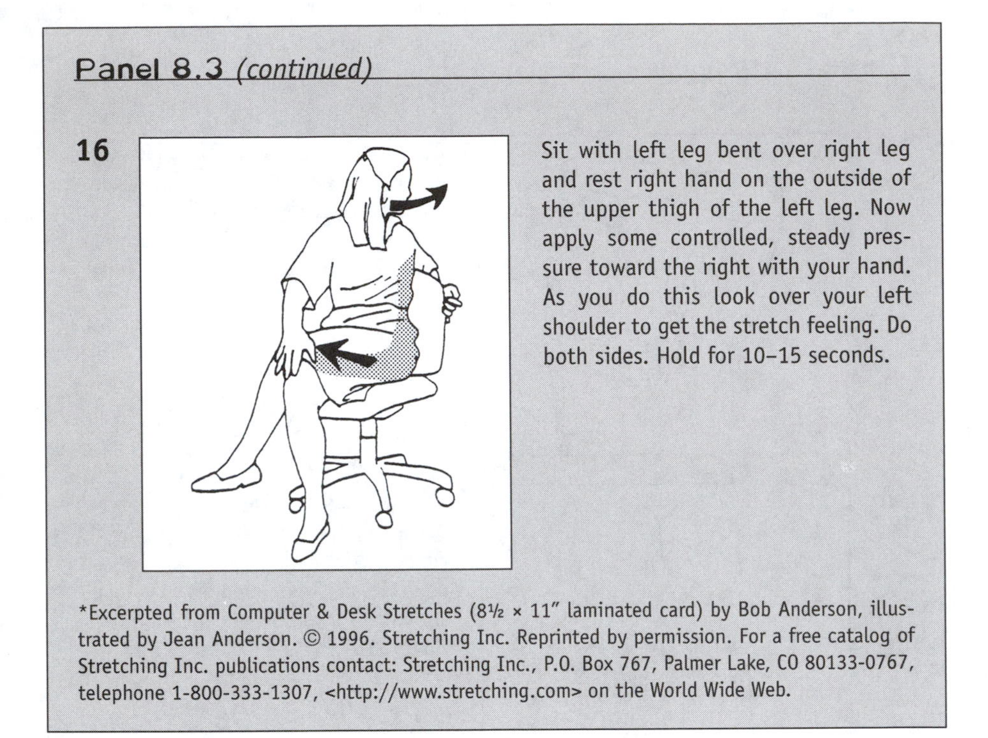

16 Sit with left leg bent over right leg and rest right hand on the outside of the upper thigh of the left leg. Now apply some controlled, steady pressure toward the right with your hand. As you do this look over your left shoulder to get the stretch feeling. Do both sides. Hold for 10–15 seconds.

*Excerpted from Computer & Desk Stretches (8½ × 11" laminated card) by Bob Anderson, illustrated by Jean Anderson. © 1996. Stretching Inc. Reprinted by permission. For a free catalog of Stretching Inc. publications contact: Stretching Inc., P.O. Box 767, Palmer Lake, CO 80133-0767, telephone 1-800-333-1307, <http://www.stretching.com> on the World Wide Web.

computer. At least every hour, stand up, walk away from your computer, or stretch to relax and loosen your muscles. Taking short breaks not only helps relax your muscles but may help clear your mind as you write.

LOOKING AHEAD

It's hard enough to write, let alone write in an environment that leads to unnecessary fatigue. Before you begin your writing project, consider our discussion in this chapter and assess your writing environment. Often, a few simple changes to office furniture, lighting, and the location of your computer monitor, keyboard, and mouse can dramatically increase you comfort level. In turn, you'll find the writing experience more pleasant than you might otherwise have imagined it could be.

chapter 9

Keeping Your Computer Healthy

To be a productive writer, you need to keep your computer working smoothly and efficiently. Like all equipment, computers break down. Problems can arise with hard disks, floppy drives, monitors, and more. Fortunately, these kinds of problems are relatively rare. Most computers are highly reliable—particularly if you take care of them. If you buy a computer from a reputable dealer and take care of it, it's likely to work well for five or more years. The key, of course, is taking care of it. In this chapter, we focus on strategies you can use to minimize problems with your computer. Our discussion covers preventive maintenance, controlling electrical power, keeping a clean workspace, keeping software working, loading and unloading software, and protecting your document files.

Our discussion applies to all computers—yours, a friend's, a company's, or a school's. If you're using a computer that belongs to someone else, treat it with care, but leave the maintenance to them. If you notice something wrong, don't try to fix it yourself. Instead, let the owner or a computer technician know about the problem in detail through a conversation, a phone call, an electronic mail message, or a note.

PREVENTIVE MAINTENANCE

Preventive maintenance is the key to keeping your computer running properly. The United States Army, Air Force, Navy, and Marines run on equipment, and all branches keep their equipment running through preventive maintenance programs. The philosophy is a simple one: take care of a piece of equipment, and it will take care of you. The key to preventive maintenance is knowing your computer.

Each computer has characteristic sounds, noises, vibrations, and even smells. Stay attuned to them, and you'll be on you way to keeping your computer, printer, and other peripherals running well. Consider how your computer sounds when you turn it on, how it sounds when you save a file to the hard disk, when you save a file to a floppy diskette, and when it carries out its other functions. When you turn on most computers, you can hear the fan on the power supply as it sucks cool air in through the computer's vents and exhausts the hot air out the back side of the computer. Take a look at the back of most computers and you'll see the power supply fan. It should be turning and blowing a steady stream of warm air out of the computer. The air flowing over the internal parts cools them, keeps them from heating up and burning out (see Figure 9.1).

In addition to the power supply fan, many 486 personal computers and all Pentium and Pentium II personal computers have a heat sink (a piece of metal to dissipate heat) and fan mounted on the CPU (the central processing unit or main chip). As the CPU carries out its silent processing, the CPU builds up heat, the heat moves to the CPU sink, and the fan carries the heat away from the CPU. The power supply fan exhausts the heat outside of the computer case. If the CPU overheats, your computer may act strangely, and eventually the CPU may burn out. If the CPU does burn out, replacing it can cost a great deal of money, depending on the CPU (see Figure 9.2).

Monitors also give off heat. Unlike computers, however, most monitors do not have a fan. Instead, they depend on the natural flow of air through the monitor vents to keep them cool. Look at the top of most monitors and you'll see a

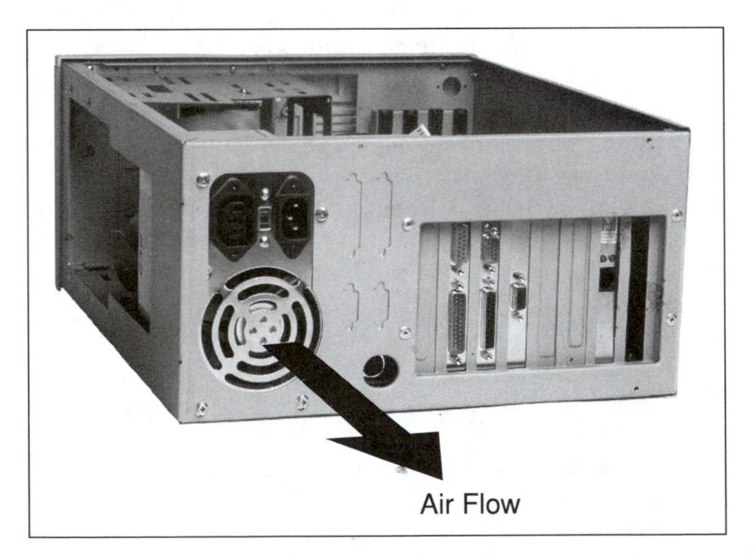

FIGURE 9.1 Close up showing the rear fan with arrow showing flow of hot air out of the back of the computer

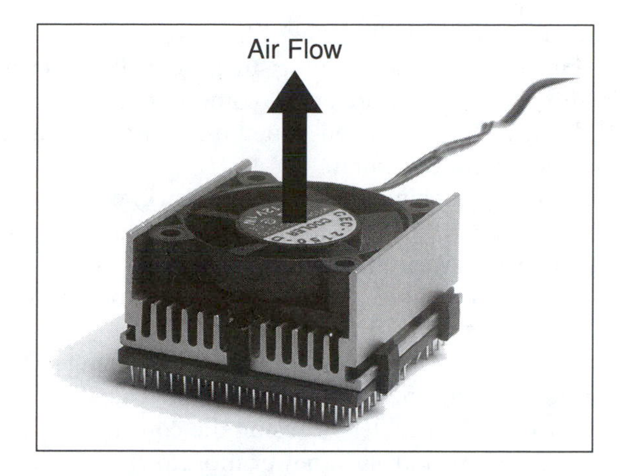

FIGURE 9.2 Closeup showing the heat sink and fan on the CPU

series of vents. Put your hands over the vents and you'll feel the heat rising from the monitor. Block that heat flow, and you risk damaging your monitor. It's a good idea, as a result, to leave space around the sides, top and back of your monitor. Don't block air flow by piling papers on top of a monitor or placing books and files along the sides.

The same goes for printers. They have fans in them that cool the internal electronic parts. Don't pile papers and books around the sides and vents on printers blocking the air flow.

On occasion, we've heard a groaning sound coming from our computers when we turn them on in the morning. Experience has taught us that it's the sound of a CPU cooling fan acting up. Usually, the groaning goes away if we turn the computer off and back on again. If the groaning persists, we've called a computer technician to replace the $15 to $25 fan. Although we haven't been happy about the need to replace the fan, we recognize that it's much cheaper than replacing a $300 CPU.

Hard drives, floppy drives, and CD-ROMS also make some noise. They spin at high speeds—in some cases at several thousand revolutions per minute. As you use your software programs, computers spin these drives, searching for the disk sectors with the needed code. Not only will disk drives make noise, but they also create small vibrations. In most cases, you'll hardly notice these vibrations, but should something begin going wrong, the drives will begin vibrating more and more. Listen to several computers and you'll soon learn the range of drive noises and vibrations. If you can put your hand on the computer and feel the vibration, have it checked.

Computers rarely have a distinctive smell about them—a noticeable smell is usually a sign that electrical parts have become excessively warm. Some people describe an electrical smell as smelling "like after a spring rain." It's the smell of ozone. Other people describe it as a burning smell. Actually, describing the smell in words is quite difficult. As you work on the computer, it's possible that you will notice an unusual smell. If you do smell something, turn your computer off and have a technician check your computer.

Spotting problems by looking at a computer works only for external parts: disk drive doors, the keyboard, cabling, the back of the computer, the mouse, and the monitor. The most likely problems you can find by looking at computers include loose or unconnected cables, broken keyboards, and broken doors on the disk drive, CD-ROM drive, or DVD drive.

As you use your monitor, pay close attention to any changes in the image on your monitor. At times the image may begin to shrink, change colors, or fade. In some cases, you can solve the problem by using the control buttons on the front of the monitor or onscreen controls in newer monitors. These controls usually include aligning the screen image on the monitor, changing the contrast, and adjusting or decreasing the brightness.

When you start your word processing programs, most have the default set with a white background and black letters. Once you have the contrast and brightness set for your use, the monitor should maintain the settings. Over time, screen image changes may signal possible problems with your monitor. Watch for color changes, gray patterns, shrinking screen size, and flicker. Sometimes the problems will clear up on their own; at other times, they signal the need for repairs or replacement.

CONTROLLING THE POWER

Although most electrical power supplies provide adequate power, the power fluctuates. These fluctuations can damage your computer, monitor, or printer. Differences in the levels of electrical power coming to your computer come from five sources:

1. The electricity being supplied to your home or office by the power company
2. The number of computers and other electrical equipment on the electrical circuit in your office or home
3. The electrical current powering the telephone system from the telephone company, if you have a modem
4. The rare, but possible, lightning strikes that hit power or telephone lines
5. Static electricity

Although the sources vary, electrical problems usually arise from:

- Too much electricity and surges
- Too little electricity
- Too much static electricity

Too Much Electricity

Electrical power fluctuations and surges, sometimes called spikes in electrical power, can damage your computer. Power failures, whether local or regional, can damage computers, monitors, printers, and other devices.

Use a good surge protector and uninterruptible power supply and you'll reduce the likelihood of having power problems. Surge protectors are electrical devices that control the amount of electricity coming into your computer. If too much comes through, the surge protector reduces the power and keeps the surge of electricity from damaging your computer. Uninterruptible power supplies include a surge protector and a battery backup system. If the electrical current coming into the computer drops below a specified level, the uninterruptible power supply sounds a warning and its battery provides enough electricity for you to save your work and shut down the computer.

When you buy a computer, invest in a good surge protector and uninterruptible power supply. Purchase a unit that provides modem protection. Such units have two telephone jacks; one for inserting the incoming telephone line from the wall, and a second for the telephone line from the surge protector to the modem. Surge protectors cost from $10 to $200, while power supplies cost $100 or more. As with most equipment, you get what you pay for.

Quality surge protectors and uninterruptible power supplies will have small lights, usually green and red, that indicate whether the device is working. Read the instructions that come with the device to learn what the different lights mean, then check the surge protector lights regularly. As surge protectors arrest surges over time, they may wear out and ultimately fail. The small lights on surge protectors signal whether they are working, are connected properly to the power supply, and whether the surge protector needs to be replaced.

If your surge protector or uninterruptible power supply fails, keep in mind that it saved your computer—the surge protector took the damage rather than the expensive machine. Although replacing the surge protector or power supply may cost $100 or more, that's less expensive than the repair bill you'll face should your computer be damaged by an electrical surge.

Too Little Electricity

For three years we struggled with computers in our research center. The computers crashed, had a range of problems, and the software became corrupted. Kelly, a new computer technician, came by to solve a problem one day and asked, "Have you had the power supply on this circuit checked?" We hadn't.

We asked a university electrician to check the power supply on the electrical circuit, and he found an overloaded circuit. The circuit had a copy machine, four computers, and two bays of video editing equipment attached. When a staff member was working on the computer, and someone turned on the photocopy machine, the computers crashed. Once we learned the source of the problem, we installed a $300 power supply with a battery and surge protector. Immediately, the computer problems ceased. Although the unit was costly, it not only saved the computer from continuous damage, it saved dozens of hours of staff members' time and eliminated our frequent calls to the computer technician.

Static Electricity

If you've ever walked across a carpeted room, touched a metal object and had a spark zap you, you've experienced static electricity. It's more common in dry climates and during the winter. In some cases, it can damage computers. Cautious computer technicians know static electricity can damage memory chips, the CPU, and other electronic parts within a computer. When they work on a computer, they wear electronic ground straps on their wrists and ground themselves before touching any electronic parts.

If you find you're working in a room where you frequently generate static electricity, ground yourself before you touch your computer. Use anti-static spray that comes with computer cleaning kits, use static grounding strips, or touch a grounded metal object before touching your computer. Check with computer supply stores for static grounding strips, wrist straps, and other devices. Obtain and use them to minimize the chances of zapping your computer with static electricity.

KEEPING A CLEAN WORKPLACE

Dust, water, soft drinks, coffee, tea, food, smoke, and other substances can easily damage your computer. Dust can build up on the vents and parts inside your computer, while liquid spills can damage keyboards and disks. Computers, like all equipment, work better and have fewer maintenance problems if you keep them clean.

Prevention is the easiest way to keep your computer clean. When using a computer, follow these suggestions to minimize potential damage:

1. Don't set coffee, tea, soft drinks, water or other liquids on your computer desk
2. Don't drink coffee, tea, soft drinks, or other liquids above the keyboard
3. Don't eat above the keyboard
4. Don't put food on your computer desk
5. Don't smoke around your computer
6. Don't place your computer near an open window
7. Don't use pencil erasers around computers

Although these guidelines appear to be simple and common sense, many writers break them on a regular basis. The solution, aside from not eating, drinking, or smoking near your computers, is to clean your computer regularly.

Washing your hands regularly is another useful strategy. Eating some foods, such as potato chips, and then typing leaves a greasy film on your keyboard. Likewise, touching your face while writing leaves oils on your fingers. As you write, your fingers transfer the oil to the keyboard and the mouse.

Although it's not always practical to wash your hands before you use a computer, doing so two to four times a day helps reduce body oils and other grime that may accumulate on your keyboard, mouse pad, and wrist rest.

Cleaning Your Desk and Computer

It does little good to clean your computer if it's on a dusty desk in a dusty room. Consider regularly vacuuming and dusting the room in which you keep your computer. Remember, though, to make sure the vacuum is not too powerful before you vacuum your computer, monitor, keyboard, or peripherals—it might damage parts. Avoid shop and industrial vacuums; they may be too powerful.

It's also a good idea to wipe down your desk using a cloth dampened with a cleaner to collect any dust, dirt or grime that vacuuming missed. *But don't use household cleaners on your computer, monitor, or peripherals—they may damage them.* Instead, purchase a cleaning kit for computers. Most computer stores stock cleaning kits for personal computers; these kits include cleaning cloths and foam-tipped probes for use on exterior cases of monitors, keyboards, printers, and mice, as well as directions for cleaning computers.

Whatever you do, don't spray any liquids on the keyboard, into disk drives, or into the vents on computers, monitors, and printers—you might damage internal parts. Instead, spray a small amount of the supplied cleaner on a cloth and then wipe the designated part.

For cleaning keyboards, place a small amount of the cleaner on a cloth and gently clean each key. As you work, change to a different spot of the cloth, add more cleaning solution, and continue cleaning the keys. Regularly cleaning keyboards works better than letting dirt build up and then trying to clean them.

Some cleaning kits contain anti-static solutions to spray on your computer. These sprays reduce static electricity, which attracts dust. Before you use them, read the directions, and follow them closely. Some cleaning kits also contain special disks and head-cleaning solutions for cleaning floppy disk and CD-ROM drives. The kits contain a cleaning floppy diskette made of a fiber. To use one of these, place two to five drops of a head-cleaning solution on the cleaning floppy diskette, insert it into the disk drive, and activate the drive. The cleaning diskette will clean the disk drive.

Mice sometimes become sticky from dirt that they accumulate as you roll them across a mouse pad. It's often oil from your fingers and dust that combine to clog the small rollers inside the mouse. You can open most mice by turning a locking device (Figure 9.3a–b). Remove the ball and then inspect the inside for dirt that may have accumulated on the small rollers (Figure 9.3c). Place a small amount

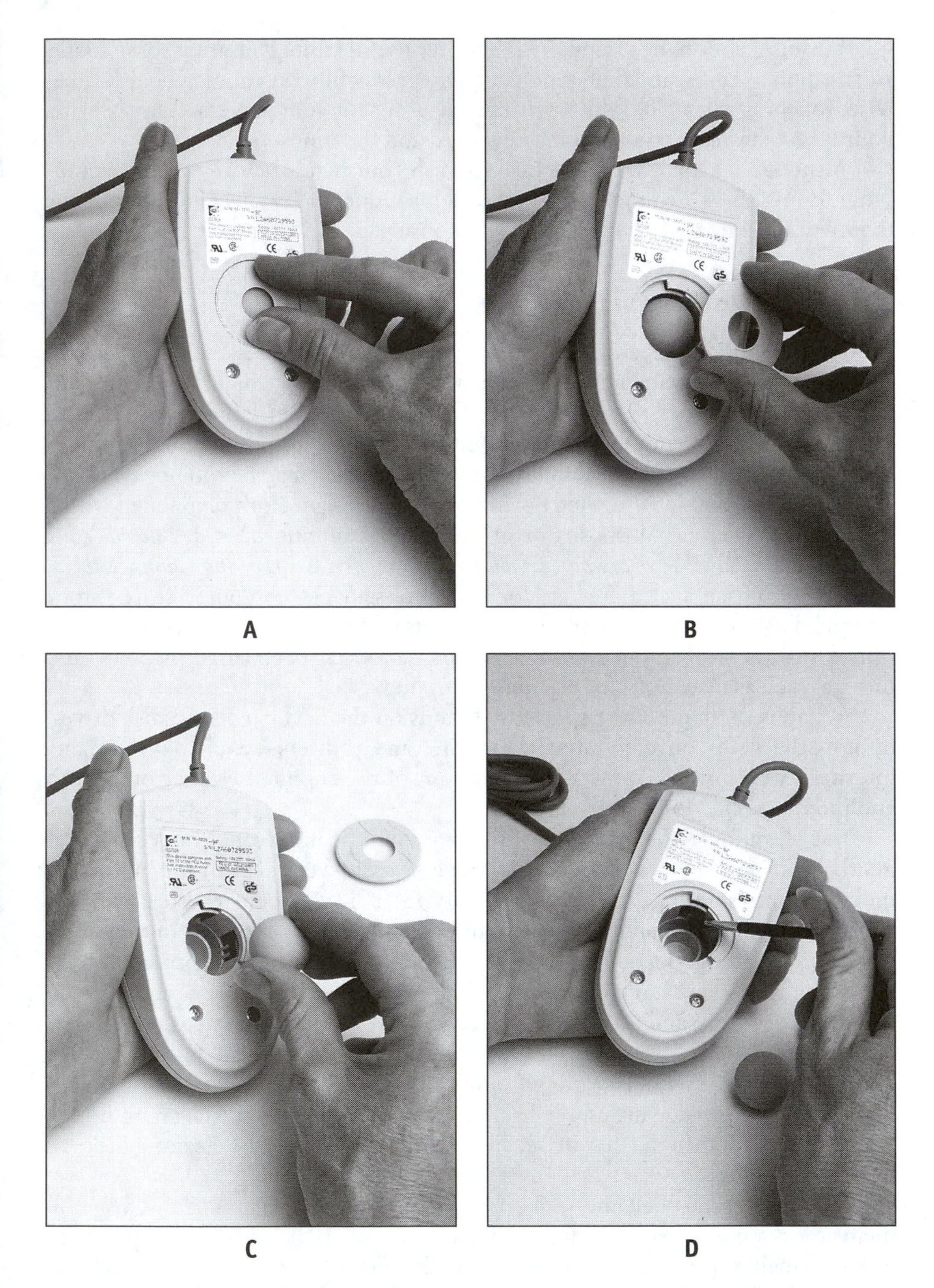

FIGURE 9.3 Close-up of sequence removing the mouse ball, seeing the rollers, and then cleaning them

of designed mouse cleaner on a foam probe and remove any dirt from the rollers (Figure 9.3d). Then wipe the rubber mouse ball with a clean, dry cloth. Some cleaning compounds can damage rubber mouse balls; use only approved cleaners.

If your mouse pad has a hard or slick surface, apply a small spot with glass cleaner to see if it might damage the surface. Wait a few minutes to make sure the cleaner doesn't soften the surface. If it's ok, clean the entire mouse pad to remove any accumulated oils and dirt so that your mouse doesn't pick them up as you use it.

KEEPING SOFTWARE WORKING

One summer, Ellen was doing a workshop in a conference room next to Reid's office. During a break, Ellen walked into Reid's office and asked Reid to print a file from her diskette. Reid agreed, inserted the flopping diskette into his computer, and his virus checking program automatically checked the diskette. It stopped suddenly, indicating that a virus was on Ellen's diskette. Reid quickly pulled the infected diskette from his computer. Ellen insisted she'd had a virus checker on her laptop and the diskette didn't contain a virus—Reid's machine was in error. Reid refused to open the file. Ellen then tried another machine only to learn that the virus checker she was using was outdated and could not catch the virus on her diskette. When Ellen upgraded her virus protection software, the new version of the checker found a virus on her laptop.

Although software is difficult to damage, problems can occur. In this section, we focus on two key strategies for keeping your software working: avoiding viruses and solving software corruption problems.

Avoiding Viruses

Viruses present one of the leading causes of damage to document files and computer software. Viruses are software codes that anti-social programmers develop to disrupt computers or damage computer software and data files. Some viruses are relatively innocuous, while others can cause major problems.

You can infect your computer by using infected floppy diskettes, downloading files from the Internet, and on rare occasions, installing virus-infected software on your computer. Viruses move from one machine to the next by hiding in document files or software programs. When you use them on your computer, you activate the virus, and it infects your hard drive. In some cases, viruses lay dormant for weeks or even months, only to attack on specified dates or under specific conditions.

To avoid infecting your computer and document files, install virus protection software on your computer and be cautious when you insert diskettes from other sources in your computer or download files from the Internet.

tip 9.1 ■ 10 Tips to Maintain Your Computer's Health and Your Sanity

Keeping your computer healthy is easy, if you follow some basic tips. Follow them and you'll have relatively few, if any, disasters. These tips include:

1. Don't eat or drink near your computer
2. Don't block the air intake vents on the sides, back, or front of your computer
3. Don't put any papers, folders, or books on the top or along the sides of your monitor or printer
4. Don't write with dirty hands
5. Vacuum your keyboard, monitor, and computer case to prevent dust buildup
6. Clean the disk drive with cleaning kits
7. Clean the keyboard
8. Use a surge protector on your computer and peripherals
9. Save regularly as you work
10. Regularly backup your hard drive and document files

Not only do you need to use a virus checker on your computer, you need to update it regularly. Programmers who develop viruses see virus checking programs as a challenge and try to devise ways to out-smart the virus checkers. Thus, they strive to create viruses that elude virus checking programs.

tip 9.2 ■ Tip to Foil Viruses

To avoid infecting your computer with a virus:

- Don't share diskettes without checking them for a virus
- Avoid using diskettes from an unknown source
- Save downloads to a floppy diskette, if possible
- Check with the diskette owner before opening a document
- Run a virus checking program before opening files
- Run a virus scan on a diskette before opening any files

Although these six tips can't guarantee your computer won't be infected with a virus, they will help minimize the chances of infecting your system with a virus. If your system ever becomes infected, virus checkers have programs for removing the virus and restoring your system. If needed, turn to a computer technician for help.

Dealing with Corrupted Software

On occasion, your word processing and other software may become corrupted for a variety of reasons, among them operator error, static electricity, and conflicts during the installation of other software. When your programs become corrupted, they fail to operate properly and can create major problems. You may face difficulties in printing and working with files, among other problems.

Whenever you encounter a problem, try to replicate the error—in other words, create it again. If you can:

1. Describe the problem in writing
2. List the commands and command sequence you used
3. Review the online help or the manual on the topic

If online help or the manual does not identify the problem, seek help from a technician or a software support line. In many cases, you may need to uninstall (i.e. delete the software from your hard drive) and then reinstall the software.

UNINSTALLING AND REINSTALLING SOFTWARE

In the uninstalling and reinstalling process, keep your wits about you and you'll find the task easy as long as you follow the instructions closely. The following discussion focuses on uninstalling programs using the original software, and uninstallation software, and then installing software.

Uninstalling Programs Using the Original Software

Current versions of most word processing and other software programs contain an uninstall function for removing the software from your hard drive. In many cases, you can uninstall software without needing to use the original installation disks. In Windows 95, for instance, you can use the Add/Remove Programs tool in the Control Panel to uninstall most software.

In some cases, however, you need to use the original installation disks to uninstall software. To do this, follow four basic steps:

1. Dig out your master software diskettes or CD-ROMs
2. Insert the first diskette in the "A" drive or the CD-ROM in the CD-ROM drive
3. Type the appropriate command
4. Follow the steps on the screen

Using Uninstallation Software

Several companies produce stand-alone uninstallation software packages, such as Quarterdeck Clean Sweep. The programs cost between $50 and $100, but can save you hours of work. You'll need to have the stand-alone uninstallation programs installed on your hard drive, so it's a good idea to purchase a copy, install it, and have it ready to use when you need it. This way, should your hard drive later become full, you need only to invoke the uninstallation software.

To use the uninstall programs, click on the appropriate program icon to open the program. The uninstall software then has a series of icons that allows you to analyze the software you want to uninstall, and determine which files are unique to the program and which files it shares with other software programs.

Warning. *Don't delete files shared with other software programs or you'll need to reinstall them.* Once the program runs the diagnostics, you can remove the software you'd like to delete from your hard disk.

Installing Software

Software installs almost effortlessly with the Macintosh OS, Windows 95, Windows 98, and Windows NT. Before you begin, however, have the serial number of your software available. In many cases you'll need to type in the serial number, your name, and sometimes your address before you begin the installation process. You'll usually find the serial number on a separate card that comes in the original software package, on one of the first two diskettes, or on the CD-ROM box.

If you're installing new software, write the serial number in your manual so that you have two handy records of the number. Store the original registration number in a separate folder where you maintain your records of software installed on your computer.

Insert the CD-ROM containing the software in your CD-ROM drive, and the software should come up on the screen. Follow the on-screen directions. Generally, you'll find it easiest to do a *typical* rather than a *custom* installation—this way you need to know less about computers and how they operate. If you know a great deal about the program you're installing, *custom* installations allow you to install selected components of the software.

If you are running Windows 95, Windows 98, or Windows NT and the installation program does not start automatically:

1. Click on the START button
2. Click on RUN
3. Type D:/setup (where D is the letter of your CD-ROM drive)

If your CD-ROM drive is a different letter, replace the D: with the appropriate drive letter. In some cases, the command to run the installation is *install* rather than *setup*. If you're unsure, check the installation booklet or manual that came with your software.

If the software you're installing comes on 3.5 inch diskettes, insert diskette number 1, and follow the directions on the floppy diskette. After your computer has installed the information from each diskette, it will stop and prompt you to insert the next diskette and click on the OK button. Pay close attention to the diskette number or names. In some cases, you'll find extra diskettes that you may not use for installing parts of the software.

Prior to delivering software on CD-ROM, major word processing programs and office suites came on 20 or more 3.5 inch diskettes, and it was not uncommon to spend 30 to 60 minutes installing the software. You'd need to insert a diskette, let the computer install the information to the hard disk, then read the prompt for the next diskette, insert it into the drive, and click on the continue installation command. Should you find yourself installing software from a series of diskettes, try to find another task you can work on while you're installing the software.

PROTECTING YOUR DOCUMENT FILES

Jim panicked; he was ready to scream. He'd written his take-home examination at home, brought it on diskette to the computer laboratory, and planned to print it just before class. When he tried to open the document file, the computer gave an error message. He tried a second computer, and then a third—all gave the same error message. His instructor saw that Jim was having a problem and tried to open other files on the diskette—all with no luck. He queried Jim. He'd been using the diskette for two years for class assignments.

Something was wrong, but neither Jim nor his instructor could determine the problem. Luckily, Jim had kept a backup file on his computer at home. His instructor asked him to go home, copy the document to a new diskette, and then come to the lab and print it. Jim did as instructed and turned the examination in by the close of the laboratory period.

To preclude the loss of your files, you'll need to take care of diskettes and back up your files.

Taking Care of Diskettes

Although diskettes generally present few problems, you may, like Jim, encounter problems on occasion. The computer may not be able to read your diskettes, and may therefore give error messages. Your diskettes may become bent or warped, become damaged, or may contain damaged files.

Floppy Diskettes. Although you may encounter the older 5.25 inch floppy diskettes on rare occasions, the 3.5 inch floppy—actually it's quite stiff—has emerged as the standard diskette for personal computers. The diskettes consist of a square plastic case, a 3⅜ inch, paper-thin circular mylar disc with a metal hub, and a metal or plastic slide that covers the diskette opening (Figure 9.4). When you insert the diskette into the computer's disk drive, the slide moves to the side exposing the circular magnetic disk, and the magnetic head in the disk drive "writes" or records information on the disk as it spins.

As you collect diskettes from different writing projects, keep them in their original boxes or purchase cases in which to store them. Use a felt-tip pen to label each diskette with such key information as:

- Project or document
- File names
- Dates
- Your name

To protect your diskettes, keep them in a proper box or carrying case. Diskettes can be damaged by:

- Heat
- Cold
- Magnetic fields
- Dropping
- Bending
- Crushing or smashing
- Dust and dirt
- Liquids
- Age

That said, a commonsense approach centers on protecting your diskettes from possible damage. Don't leave diskettes where they can become excessively hot or cold—as might happen if they are left laying on the dash of a car during the summer or winter. Similarly, don't leave diskettes where they can be soaked by coffee, water, soft drinks or other liquids that might spill on a desk or computer table.

Keep your diskettes away from telephones, electrical lines, electric clocks, printers, other electrical appliances, metal, and anything magnetized as a magnetic field might scramble your files. And don't play Frisbee with diskettes. Although it's a diversion from writing, dropping the diskette might crack or break the case or bend the slide.

Don't expect diskettes to last for years, or even months. Like all equipment, they wear out. Diskettes are relatively inexpensive; by the late 1990s, careful shoppers could find good quality diskettes for less than $20 for a box of 25.

Warning. *If you know you have a damaged diskette, never put it in the disk drive. It might jam in the drive and you won't be able to remove it.* If you jam a diskette in a

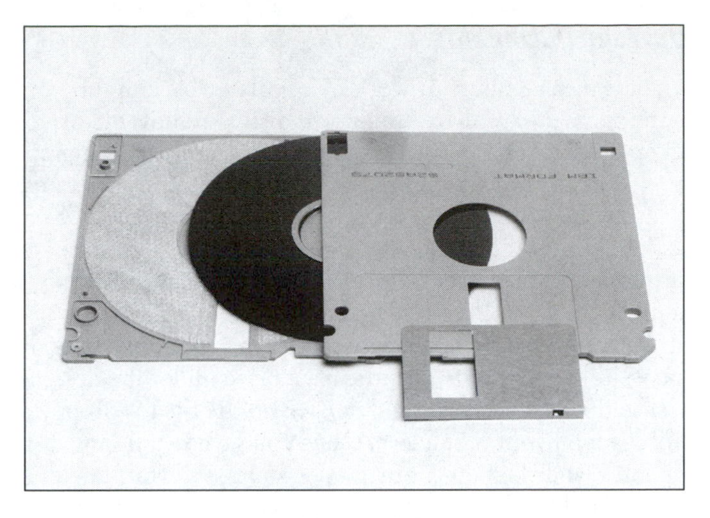

FIGURE 9.4 Dissected floppy disk

disk drive, you'll need to have a technician remove it; the charges might run between $50 to $100 depending on the time required and the costs of replacing the disk drive.

Hard Drives. Hard drives are remarkably reliable and will typically run for years before they fail—but they do fail. Hard drives spin at several thousand revolutions per minute, and those revolutions ultimately take their toll. When your hard drive fails, you're faced with the options of replacing the hard drive on your own or having a technician replace it, reload all of your software, and try to restore your document files. Restoring the files is not a problem if you've regularly backed up your work, but difficult or impossible if you haven't. To avoid losing your work, always make backups of your files.

Maintaining a hard drive is simple. Install a surge protector and clean around your computer to keep the dust out of the computer.

Hard drives record or write information to different sectors, or sections, of your hard drive as you save files. In some cases, the files are fragmented—separated into several different locations—on your hard drive. As you save more and more files to your hard drive, your computer stores the file fragments in different locations. When your computer begins searching for the fragments, it may slow down as your disk becomes full.

Computer experts suggest *defragging* your hard disks to solve the problem. Even if you're new to computers, you should install and learn how to use a software utility program that can *defragment* your hard drive.

Backing Up Your Documents

As a writer, you can save hours of work by regularly backing up your files to a floppy diskette, second hard drive, an optical disk, a removable drive, a tape, or another backup media. You can use different strategies for backing up your files.

Within a given writing session, major word processing programs allow you to schedule automatic saves of your documents. Most word processing programs come with a default setting, often every 10 minutes, but you can set the time for more or less frequent saves. You can specify options for automatically saving files in the Options, Settings, or Preferences dialogue box.

In addition to automatic backup, you should save backup copies of your document as you write. Save your files to the hard drive under the designated folder or file. Then use the SAVE AS command to back up the file to a floppy diskette, or to another location on your hard drive. Once you save to a floppy diskette, leave it in the computer, and then do another SAVE AS to save your document back to the original location. Depending on your sequence, the computer may ask whether you want to replace an existing file of the same name.

If you're sure the current copy is the latest version, then save over the older file. If you do, the computer copies the new file over the older file of the same name. Should you later decide that you'd like to use the older file, you won't be able to access it. If you're unsure, save the file under a different name in order to maintain the earlier file. You can always move it to a new folder or erase it later.

Some writers, rather than saving the files to a floppy drive as they work, save it to only the hard drive. Once they've completed writing the document, they copy their files to a floppy diskette. It's usually a faster way than using a SAVE AS command from the word processing program. But it's not quite as safe. Should you lose the file or your hard drive in some way becomes corrupted, you'll need to start over.

Beyond backing up individual files, your can use tape drive systems, removable disks, and optical disk systems to back up selected files, individual folders, or entire hard disks. Depending on the medium, you can use standard file management software or dedicated backup software. If your computer is equipped with a tape drive, removable disk, or optical disk system, review its online help and manual to learn the specifics. If disaster strikes—and it will from time to time—you'll have the needed backups to quickly recover.

Recovering from Disasters. When disaster strikes, overcoming the problem will depend on the backup system that you're using. The tasks include:

1. Locating the backup file or files
2. Opening the file or files
3. Saving back to the hard drive

Although the overall process is similar across operating systems and backup software, you'll use different steps depending on whether you backed up the files on hard disks, floppy diskettes, optical disks, tapes, or removable disks. Because of the differences across media, we'll focus here only on floppy and optical disks. For other media, check the manuals and instructions that accompany the hardware and software.

Assume you were working on the word processing program and for some reason lost your file. First, check the local folders where you were working and where you save backup files, and then search the hard drive. You can visually scan the file names on the hard drive by looking through folder after folder, but that's a time consuming process. To locate the files more quickly, use the operating system's file management software to search the entire hard drive or selected folders on the hard drive.

Once you locate the file, copy it to the desired folder or open it in the word processing program and then use the SAVE AS command to save it to the desired location. Once you've saved the file, read down through it to learn the last changes. In many cases, you may have lost only a paragraph or two. In other cases, you may find it necessary to rewrite several pages—all depending on how often you'd been saving files as you worked.

LOOKING AHEAD

Poorly maintained computers have the annoying—but predictable—habit of breaking down just when you need them most, usually when you're working on deadline. To avoid unnecessary problems, consider this chapter. You'll find that an ounce of prevention is truly worth a pound of cure—or, in this case, the time it takes to recover a lost file, disinfect a computer virus, or reinstall a software program.

chapter 10

When Disasters Strike—Solving Computer Problems

No matter how skilled you are when using a computer, disasters do strike—the electricity dies suddenly, you see an error message that says "Disk full," a virus strikes, your monitor turns strange colors, your computer groans when you turn it on, your hard drive crashes, or any one of a dozen other problems come up. When disasters strike, you can be up and running sooner if you use troubleshooting strategies and know how to solve common problems. In this chapter, we discuss a basic troubleshooting strategy and take a look at 11 common problems.

AN EXAMPLE—STEVE'S COMPUTER WON'T PRINT

About 8 a.m., Steve turned on his computer and printer, opened his word processor, wrote for 30 minutes on a document, and clicked on the PRINT icon on his button bar. After two minutes nothing had happened. Rather than panic, he began troubleshooting his problem. The afternoon of the day before, everything had worked fine.

Steve knew that computers and software interact together in complex and sometimes mysterious ways. For starters, Steve knew that the problem could be an operator error—perhaps he'd failed to do something correctly. He also knew it could be a problem with his word processor, his operating system, his printer, his printer cable or connector, or the hardware in his computer. To begin solving the problem, Steve asked himself:

- Is the printer on?
- Is the printer *online* button turned on?
- Is there paper in the tray?

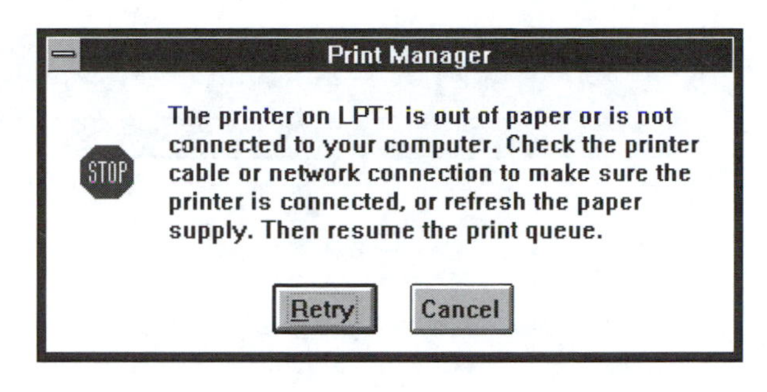

FIGURE 10.1 Printer error message

- Is the paper inserted correctly?
- Is there a paper jam?

His printer was on, the *online* button was on, paper was inserted correctly in the tray, and no paper jammed the printer. Steve began checking other possible causes:

- Had the printer signaled a problem?
- Had the word processor signaled a problem?

Steve checked the printer's digital readout window—no problems were indicated. Further, Steve knew that the printer, computer, and word processor often signaled a printing problem through the Print Manager (Figure 10.1). The Print Manager window suggests possible problems: it will warn you if the printer is out of paper, if a cable connection problem exists, or if a network problem exists.

In following up on the next questions:

- Was the paper supply adequate?
- Were the cables connected tightly to both the printer and computer?

Steve checked the paper supply and the printer cables. The paper supply was OK. The cable connections were tight. To be sure, he gently pushed the connectors again. Still the problem persisted.

- Had the print manager software received the file for printing?

Steve minimized his word processor, looked for the Print Manager icon (Figure 10.2), and clicked on it. The Print Manager window opened up (Figure 10.3); no print jobs were waiting.

Steve still couldn't solve his printing problem. He asked himself other questions:

FIGURE 10.2 Printer icon

- Is it the word processing software?
- Is it the printer control software?
- Is it a more serious hardware or software problem?

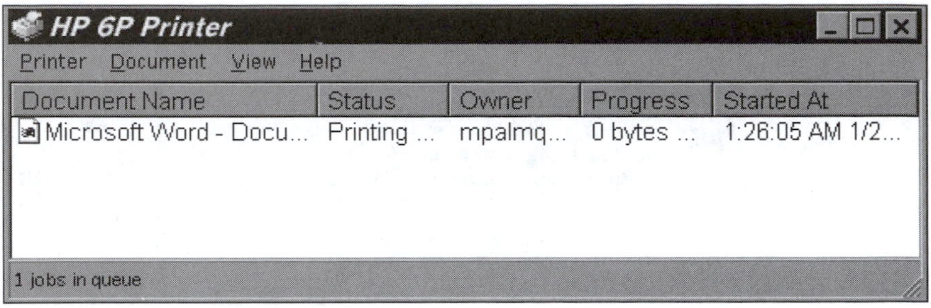

HP 6P Printer

Printer Document View Help

Document Name	Status	Owner	Progress	Started At
Microsoft Word - Docu...	Printing ...	mpalmq...	0 bytes ...	1:26:05 AM 1/2...

1 jobs in queue

FIGURE 10.3 Printer manager window

Steve opened his electronic mail software and tried to print an electronic mail message. The electronic mail disappeared, but did not print. Steve clicked on the print manager icon, and reviewed the print manager window. No files were waiting to be printed. Steve couldn't figure out the problem, so he saved his file again, closed down the word processor, closed down the operating system, and turned off his computer. He waited a few seconds, turned on his computer and printer again, opened the operating system, opened the word processor, opened the file, and tried to print again. This time, everything worked and the file printed.

As you gain more expertise with computers, you'll be able to solve roadblocks like Steve's printer problem. When faced with a difficulty, go about solving it in a systematic and organized way. The following discussion provides a strategy that will allow you to solve many problems: look for the simplest solutions first.

TROUBLESHOOTING COMPUTER PROBLEMS

When a disaster strikes, don't panic; you can solve most problems by following a troubleshooting strategy. Troubleshooting moves from the simple to the complex based on yes or no questions. If you answer "no," you've eliminated a series of problems. If you answer "yes," you've narrowed your focus to a limited number of possibilities.

In this section, we warn you about possible electrical dangers associated with troubleshooting computer problems and about the complications of organizational policies, identify needed supplies and equipment, and suggest a six-step troubleshooting strategy.

Warning. Before you proceed, remember that computers are powered by electricity, as are monitors, printers, and peripherals. Don't take the case or cover off a computer, printer, monitor, or other equipment and explore its internal workings unless you've turned the computer off and unplugged it. Left on, you could receive a serious, potentially lethal shock.

Warning. Some organizations draw clear lines of responsibility for maintaining and repairing computer equipment. If you're a student in a lab and the computer you're using starts having problems, think twice before you start trying to solve the problem. The lab's student monitors can let you know what the policies are regarding trying to solve computer problems. Similarly, if you are an employee who is having problems with a computer, check before you start working on a problem. The computer may be leased, and thus its maintenance may be the responsibility of the leasing company. Or your company's information services group may have established policies governing the repair and maintenance of computers and related equipment.

Troubleshooting Supplies and Equipment

You can solve many problems yourself if you think through the problem and have the necessary equipment and supplies. You'll need:

- A basic tool kit
- A mouse-cleaning kit
- A computer-cleaning kit
- Master software disks
- Manuals and instructions

Check most computer stores, office supply stores, and computer catalogs and you'll find basic tool kits that contain standard and philips head screwdrivers, Torx (star-shaped) screwdrivers, chip extractors and inserters, nut drivers, tweezers, and a three-pronged parts retriever. Basic tool kits generally cost between $10 and $25. Although they're not always needed to repair a computer, they do provide standard tools for tightening screws, hex nuts, and picking up screws and small parts.

In addition to a tool kit, purchase a basic computer-cleaning kit and a mouse-cleaning kit. Computer-cleaning kits usually contain a cleaning diskette, diskette head cleaning solution, wipes, foam-tipped probes, video-cleaning solutions, a case and cabinet cleaner, and related supplies. Some kits contain mini-vacuums, but most mini-vacuums lack sufficient power to be of any real use. Mouse-cleaning kits include a cleaning solution, pad, cleaning balls, and instructions.

Once you have installed software on your computer, store your diskettes and CD-ROMs in a safe place, but have them readily available should you need them. If your computer came with the software installed, your should have received the master software diskettes or CD-ROMs. If not, ask the company from which you purchased your computer for a master set of diskettes. You can't troubleshoot some software problems without the diskettes.

Keep all software and hardware manuals handy. You may need them to check for specific problems and for technical support telephone numbers. Some software and hardware companies provide toll free telephone numbers for technical support, while others have standard long-distance numbers; some provide unlimited support, while others charge for support. Many software and hardware companies provide Web sites, fax numbers, bulletin boards, and other online support. Scan your manuals to see what resources companies provide for your software and hardware.

A Six-Step Troubleshooting Strategy

Think of troubleshooting as a process of elimination. Check for simple problems before you assume it's a complex problem. Consider the following problem-solving strategy (Figure 10.4):

1. Look for simple solutions
2. Try to replicate the problem
3. Seek help
4. Uninstall the software
5. Reinstall the software
6. Seek technical help

Below, we consider each step in detail.

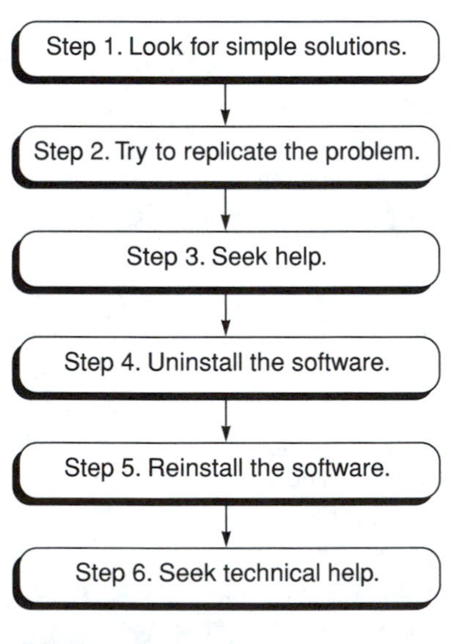

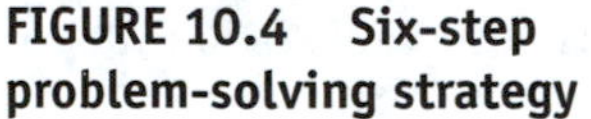
FIGURE 10.4 Six-step problem-solving strategy

Step 1: Look for Simple Solutions. You can solve many computer problems yourself. When a problem arises with a computer, look for obvious problems before you seek help from others. First, try to determine whether you have a hardware or software problem. Here's a list of 23 questions that reveal common hardware problems:

1. Are all of the power cables connected? Computer? Printer? Monitor?
2. Is the surge protector on?
3. Is the computer turned on?
4. Is the monitor turned on?
5. Is the printer turned on?
6. Is there power to the electrical wall socket?
7. Are all cables connected?
8. Does the computer cycle through the operating system and load?

9. Does the word processing software load properly?
10. Is the floppy diskette dirty? Broken?
11. Is the floppy diskette seated in the drive?
12. Is the floppy diskette door closed?
13. Is the CD-ROM disk properly seated in its drive?
14. Is the CD-ROM disk drive door closed?
15. Is there paper in the printer?
16. Is the paper loaded properly?
17. Is the toner or ink cartridge functional?
18. Is your computer connected to a network?
19. Is the printer *online* button on?
20. Can you save files to the hard disk?
21. Can you save files to the floppy diskette?
22. Is the network slow to respond (if your computer is connected to a network)?
23. Will the printer print from other software programs, such as a graphics editor or a Web browser?

The last section of this chapter identifies common problems by categories and suggests simple problem-solving strategies. Keep in mind that you may encounter many more problems when using a computer.

Step 2: Try to Replicate the Problem. When you encounter a problem, try to replicate or repeat it. To help isolate the problem:

- Write down the steps you took
- Write down any error messages the computer displayed
- Write down the time and date of the problem
- Keep your notes near your computer

By replicating the sequence that leads to the problem and taking notes, you have begun the troubleshooting process. Note especially any error messages that your computer displays. They usually provide clues that will help you solve problems.

By replicating the problem, you may find the problem to be an operator error—in other words, you're making a mistake as you type commands. In many cases, you may learn that you've been making a simple error. In such cases, you need only follow the online help or printed manual to solve the problem.

If you can't solve the problem and you need to seek additional help, you have a record of what occurred, what you did, the problem that occurred, and any error messages or computer actions that followed. When you seek help, you'll have the needed details to help solve your problem. You'll be able to fully describe the difficulty, the sequence of events that occurred prior to the difficulty, and any troubleshooting efforts you've undertaken to solve it.

Step 3: Seek Help. When trying to solve problems, first seek help by:

- Checking online help
- Checking printed manuals, if available
- Asking colleagues and friends for help
- Calling local technical support
- Calling company support lines

How you describe the problem may not be the same way it is described in the online help and printed manual. Further, computer technicians may describe the problem in different terms than you're using. Thus, you'll need to try different terms and words when searching for information. Further complicating your task is the inconsistency in terms among software programs manufactured by different companies.

Using Online Help. When you're trying to solve software problems, first check to make sure that you're following the right commands to carry out the activity you're trying to accomplish. Double-check the online help by searching for the function and reading the instructions. If necessary, print the instructions to ensure that you're carrying out the task properly.

Using Manuals. If you cannot solve the problem by using online help, turn to printed manuals. Look in the index and table of contents for key chapters related to the general problem. Many manuals are well written and provide extensive details, but finding that information can be a challenge. If you're new to using a particular computer manual, take time to familiarize yourself with the fundamental sections, how it's organized, and its index and terminology.

Asking Friends and Colleagues. As computers have become more and more commonly used, colleagues and friends often turn to one another when trying to solve computer problems. As you and your colleagues gain more and more experience with computers, they too will become a good source for troubleshooting problems. One of the key advantages of turning to colleagues and friends is that many have experienced similar problems and may be able to quickly troubleshoot your problem, as long as you describe it properly.

Calling Local Technical Support. If you're working in a university, college, business, or industry, many organizations have computer staff technicians that provide technical support. Many university and college laboratories have laboratory assistants who can solve simple problems, as well as computer support staff who can solve more complex problems.

Calling Software and Hardware Technical Support. Most companies have technicians that you can contact by telephone, fax, or electronic mail. Many companies also maintain Web sites that provide troubleshooting guidelines for a wide range of problems that customers have encountered. You'll find the telephone and fax numbers, Web sites, and electronic mail addresses in the online help, installation manuals, and reference manuals.

When calling a technical support line, keep in mind that providing technical support is expensive. When all costs are totaled, each support call typically costs the hardware or software company between $25 and $100. When you call, be prepared—have your notes ready, your computer on, and be ready to explain the problem in detail. Being prepared will save you time and save the company money.

Step 4: Uninstall the Software. At times, support technicians will recommend reinstalling the software. In many cases, you can only solve the problem by first uninstalling the software. You can uninstall software in three ways. First, if you're using Windows 95, Windows 98, or Windows NT, you may be able to uninstall the software from the Add/Remove Programs tool in the Control Panel.

Second, you may be able to uninstall the software by clicking on an uninstall icon in the software program group or folder. Or you may be able to uninstall the software by clicking on the install program on the original installations disk or diskettes. Check your installation manual about how to uninstall the program.

Third, you can use uninstall software that's designed for removing software from your hard drive. The programs usually run between $50 and $100, and you'll need them installed on your hard drive. Should you then need to uninstall a specific program, you click on the uninstall software icon, and follow the online instructions. Some programs enable you to make a backup of the software. Should you uninstall the wrong software or portions used by other software programs, you can easily recover from your mistake.

Uninstalling software cleans the system and allows you to start over. If you skip the uninstall step and merely reinstall the software, the programs sometimes skip reinstalling some of the files that you need to replace to eliminate your problem.

Step 5: Reinstall the Software. To reinstall your software, you'll need your master diskettes or CD-ROM. Simply install the software as if you were installing it for the first time. When reinstalling your software, you'll need to decide on a typical or custom installation. If you have little experience installing software and your hard drive has plenty of space, do a typical installation.

Step 6: Seek Technical Help. If you can't solve the problem with the five steps detailed above, return your computer to a service center or have a service technician perform on-site repairs. If your computer is under warranty, check your warranty. Some warranties provide on-site service while others require you to deliver or ship the computer to a service center.

If your computer has an on-site warranty, call the service center and make an appointment for the service technician to come to your home or office to work on the computer. If your computer has service center warranty, call the technical support staff, explain the problem in detail, explain how you've tried to solve the problem, and then ask for their guidance. In some cases, technicians can suggest additional troubleshooting methods to solve the problem. If that doesn't work, they'll suggest returning your computer for servicing.

If you must carry the computer in for repair, insert a diskette in the disk drive, carefully pack the computer, and place it on the floor of the car. If you must place a computer on a car seat, secure it so that it does not fall off. Use a car safety belt to hold the computer in place. When carrying computers, keep in mind that they're often off-balanced—in other words, the corner of the computer with the power supply usually weighs more than the other parts of the computer. You'll feel the difference when you pick up the computer. Grasp your computer so that you have a solid grip, and carry it close to your body so that it's well balanced. If possible, have someone open doors for you.

If you must ship the computer in for repair, use the original box and foam pads that protected the computer. If you've recycled them or thrown them away, try to obtain another large box and packing adequate to protect the computer. Use heavy packing tape so that the shipping box does not break open. When you ship a computer, insure it in case it is damaged or lost during shipping.

When you deliver or ship a computer for repair, be prepared for possible delays. Although some service centers provide 24-hour turn-around, others may take a week or longer because of a backlog of computers to be repaired, delays in ordering parts, or other problems.

COMMON HARDWARE PROBLEMS

The following discussion identifies common problems you might encounter and suggests strategies for solving each problem. Keep in mind that the following discussion is by no means all inclusive. It does, however, identify problems that we have encountered over the last 15 years when working with computers. For each problem, we suggest simple solutions that can often be used to solve it.

1. *The Computer Won't Turn On*

- Check the on/off switch
- Check the power cord
- Check the surge protector
- Check the power to wall outlet

Work from the computer to the power supply checking for possible problems. Try the on/off switch again. If it appears to work but the computer does not come on, check to make sure that the power cord is securely plugged into the computer and the surge protector (or power supply). If it is, check the surge protector or power supply to see that it's plugged into the wall, turned on, and working. Check the monitor lights on the surge protector. Push the reset buttons on the surge protector. If they're off, check the electrical outlet to make sure that you have electrical power to the wall plug. Plug a desk lamp in and see if it lights. If not, check the circuit breakers controlling power to the electrical outlet.

2. The Diskette Won't Go in the Disk Drive

- Inspect the diskette
- Inspect the disk drive orientation
- Inspect the disk drive door
- Inspect the disk drive

First, look at the diskette. Is it bent, wrapped, twisted, chipped, broken, or in some other way damaged? Check the slide. Does it move back and forth with ease? Check to make sure that you have the diskette oriented in the correct position to insert into the disk drive. To insert diskettes, you'll need to rotate the diskette. If you have the orientation correct, check to make sure there isn't a diskette already in the disk drive. Check the disk drive door to see that it opens and closes. They sometimes break on their hinges.

Warning. *Never place a damaged disk into a disk drive. You may damage the disk drive and that might necessitate replacing it—an expense of easily $75 or more.*

3. Can't Remove the Disk from the Disk Drive

- Turn off the computer
- Lift the disk drive door or push the release button
- Try to determine what's jamming the disk drive

Sometimes disks may be slightly warped. They will drop in the disk drive only to jam when you try to remove them. Try the button or lever on the drive a couple of times to release the diskette. If it releases, carefully inspect the diskette to see if it's damaged.

Warning. *Turn off the power to the computer and unplug the computer before you proceed further.*

If the diskette is still jammed in the computer, carefully lift the disk drive door and try to remove the diskette. If you can lift the door, but the diskette doesn't come out, take the tweezers from the tool kit, try to grasp the diskette, and pull it from the drive. Don't touch any of the internal parts of the drive. If you can't remove the diskette, call a computer technician. The technician will probably replace the disk drive.

4. *The Computer Won't Read or Open a File*

- Try to open the file again with your word processor
- Try to read other files on the disk
- Try to read the file using the operating system's file management program

When your computer can't open a file, several problems may be occurring. If it's on a floppy diskette, the problem might be a bad diskette or a corrupted file. If it's on the hard drive, the file could be corrupted or damaged or the drive might be damaged. First, try to open the file again with the word processor. If that does not work, try to open other files on the same disk with your word processor.

If the word processor can't read them, use the Program Manager (Windows 3.1), Explorer (Windows 95 or Windows NT), or Finder (Macintosh OS). Click on the file. Doing so may launch your word processor and open the file. Was the file created on another computer? If so, return the diskette to that computer and try to open the file using the original software and computer. Although rare, the read and write head of disk drives may slip out of alignment. Returning to the original computer may enable you to open the file and print it. If you can't open the file, use a software utility program such as Symantec Norton Utilities to diagnose the problem and fix the damaged files.

5. *The Monitor Doesn't Work Properly*

- The monitor won't come on
- The monitor on-light comes on, but no image appears
- The image changes shape
- The image seems too dull or too bright

Turn on the computer, let it begin booting up, then turn on the monitor again. If that doesn't solve the problem, turn off the computer and monitor. Check the power cord for both the monitor and the computer. Are they plugged into your surge protector? Check the monitor to computer connection cable. Are the monitor cable and connector seated tightly in the back of the computer? If the connector has screws, use the small screw driver from the tool kit to tighten the screws. If the connector has thumb screws, check them. Tighten them, if necessary, but don't over-tighten them. Try the monitor again. If it doesn't work, call a technician.

If the monitor light comes on (it's usually a pin-head sized light near the on/off button), but the monitor doesn't come on, call a computer technician. If the image changes, try to adjust it using the vertical and horizontal buttons or on-screen controls. On most monitors, the controls are aligned horizontally just below the screen. If the image seems too dull or too bright, try adjusting it with the brightness control.

6. The Printer Won't Print

- The printer won't come on
- The printer won't print
- The printer jams

Check the printer to make sure that the power cables are connected to the surge protector and the printer cable is connected to both the computer and the printer. Make sure that the connectors are securely seated. Make sure you have electrical power to the surge protector. Turn the printer off and on again.

If you have power, but the printer won't print, check the *online* button. Press it to reset the printer. If it still doesn't print, check the printing software on your operating system. Check to see if the computer and printer are "talking" to one another—in other words, sending signals back and forth. Check for possible paper jams in the computers. Although some printers have a display panel, others do not. If the printer has a display panel, check for the display message. If it indicates a jam, turn off the printer and troubleshoot for jams, as described later.

7. The Printer Prints Light Text or Streaks

- The printer prints light text
- The printer prints streaks in text
- There is a patchy appearance to the printed text

These symptoms often signal problems with the toner cartridge in laser printers, ink cartridges in inkjet printers, or ribbons in dot matrix or impact printers. Replacing the toner cartridge, ink cartridge, or ribbon usually solves the streaking and light text. Be aware, however, that some refurbished or recycled cartridges and ribbons may not produce quality as high as the original manufacturer's cartridges and ribbons. If replacing the cartridges does not solve the problem, then have a technician check your printer.

8. The Printer Groans when Printing

- The printer groans as it feeds
- The printer groans as paper moves through it

Some printers naturally make noise and others don't. Learn how your printer normally sounds and note any serious changes. Check the paper path to make sure that it's clear of pieces of paper or other obstructions—paper clips, rubber bands, ear rings, and so forth—that may have dropped into the printer. Carefully vacuum the printer to remove dust from the rollers and paper path.

Clean with printer cleaning kits. If the printer continues groaning, call a computer technician or take the printer in for a service check.

Warning. *Don't use a vacuum that's too strong or brush the internal printer parts with a vacuum brush or wand.* You could damage your printer.

9. The Paper Jams in the Printer

- The paper jams
- The paper tears
- The paper feeds at angles

Remove the paper from the paper tray or bin, check the paper path for possible bits and pieces of paper, and vacuum the printer and paper path.

Check the paper to make sure that it is not damaged, bent, or improperly cut. Try a new stack of paper. If necessary, open a new ream of paper and try it. Make sure you insert the paper properly and don't overload the paper tray or bin. Most printers limit the number of sheets of paper.

10. The Mouse Is Jerky or Won't Work

- The mouse seems jerky
- The mouse doesn't glide smoothly
- The mouse doesn't work

Over time, your mouse accumulates oil, grime, and dirt on the ball and transfers that dirt to small rollers inside of the mouse. Obtain a mouse-cleaning kit from a computer store or office supply store and clean your mouse. Clean the bottom of the mouse around the rollers. Dirt accumulates on the plastic glides. Clean your mouse and mouse pad, if you're using one. If not, buy one.

Check the mouse cable and connector. Tighten the connector screws to make sure that the connections work. As mice age, the cable may stiffen so that movement may become a problem. If necessary, replace your mouse.

Warning. *Some mouse cleaning solutions can damage the mouse balls. Never use the cleaning solution on rubber mouse balls.*

11. The Keyboard Is Having Problems

- The keyboard becomes dirty
- The keystrokes don't show on the screen
- The keys stick

Although keyboard problems do occasionally occur, they are rare. Keep the keyboard clean by wiping down the keys from time to time. Use the cleaning solutions provided in the computer cleaning kits.

Turn off your computer, spray a small amount of cleaner on a cloth and them gently wipe the keys and keyboard. Check the keyboard connection to the computer. Make sure it's seated properly, but don't put too much pressure on the keyboard plug and receptacle. If that doesn't work, replace the keyboard. In many cases, you can replace a keyboard for less than a service charge to check the computer.

Warning. *Do not spray cleaners on the keys and the keyboard.*

LOOKING AHEAD

Even the best maintained computers can develop problems. If disaster strikes, you can reduce the time it takes to get your computer up and running again by following the advice we provide in this chapter. The basic troubleshooting strategy we discuss is a useful beginning, as is the ability to recognize common hardware and software problems. In addition, talk to other writers about the problems they've encountered. Knowing what to expect can reduce the time needed to resolve your technical problems. And with those problems out of the way, you'll have more time to spend on your writing.

appendix A

Glossary of Terms

Terminology is one of the more enduring problems associated with using computers. In this glossary, we define a number of common terms.

Bibliographic Software If you write documents that include literature reviews, a bibliographic software program can come in handy for keeping track of your references and automatically generating formatted citations for your documents. Commonly available programs include EndNote, Citation, Reference Manager, and ProCite. By entering the author(s), title, date, publishers, and other information about a book, magazine article, journal article, or other publication, you create the bibliographic database. When you need to conform to a particular reference style, such as the Modem Language Association, American Psychological Association, or Council of Biology Editors, you can use the software to format your citations automatically. You need not worry about many of the minutiae of formatting citations to meet particular publication styles.

Browsers Software packages, such as Netscape Navigator and Microsoft Internet Explorer, that enable you to view information on the World Wide Web. America Online, CompuServe, Prodigy, and a host of local Internet Access Providers enable you to connect to the World Wide Web. Browsers interpret Web documents and present them on your computer. Browsers are much like word processing programs. They allow you to read files, save those files on your computer, and copy text and other elements on a page. Unlike word processing programs, they also allow you to use your modem or network card to connect to and read files on other computers.

Byte A computer unit of measure that represents one character or letter.

CD-ROM CD-ROM stands for compact disk, read only memory. CD-ROMs are optical disks—in other words, the information stored on CD-ROMs consists of small pits or indentations burned in the CD-ROM's surface. CD-ROMs can store about 650 megabytes of information. Because of their convenience, low production costs, and ease of installation, many software manufacturers distribute software on CD-ROMs rather than on 3.5 inch diskettes. For example, one CD-ROM containing an office suite (i.e., word processing, presentation, spreadsheet software, and three other programs) replaced more than 20 3.5 inch floppy diskettes.

Communication Software Software programs that allow you to connect to other computers via telephone lines or direct lines. Generally, communication software allows you to access other computers and sometimes the Internet. Special communication software in the form of browsers such as Netscape Navigator and Microsoft Internet Explorer are needed to access the Internet and use the World Wide Web.

CPU The central processing unit of computers. It is a chip containing millions of transistors that process electronic signals. Many DOS and Windows personal computers are referred to generically using the names of their CPU chips—Pentiums, 486s, and 386s, while Macintosh computers are referred to by model numbers rather than their chips, for example, 7500 and 9500.

Database Databases provide information about specific subjects. Databases typically focus on either numeric or textual information. For instance, a database might contain numeric information gained from the most recent United States Census.

DOS DOS stands for disk operating system. DOS provides the basic instructions that enable some IBM personal computers and IBM clones to operate. DOS runs in the background while other DOS-based software runs. While Microsoft's Windows 3.1 runs on the DOS platform, Windows 95 and Windows NT 4.0 have their own operating systems.

DVD-ROM As technology has advanced, computer manufacturers have developed the digital video disks or digital versatile disks read only memory disk to replace CD-ROMs. DVD-ROMs can store about 4.7 gigabytes of information, or about as much information as can be stored on six CD-ROMs. DVD-ROMs can provide full motion videos, movies, and advanced multimedia programs.

Electronic Mail Address Your electronic mail address usually consists of a variation of your name. Your network systems administrator will assign you an electronic mail account and address and ask you to provide a password so that only you can access your electronic mail account. Depending on the network and how the systems administrator sets it up, you may need to log onto the network when you turn your computer on. In some cases, your computer logs you on to the network system automatically.

F-keys Functions keys (or F-keys), usually located across the top of the keyboard, provide sequence commands to the software. They were first used on early mainframe computers; personal computer manufacturers added them to early desktop computer keyboards and they have become a fixture on most keyboards. Keyboards provide the primary way of entering information or writing on most computers.

Floppy Diskettes The floppy diskette is a magnetic disk encased in plastic. The 3.5 inch, high density diskette emerged as the industry standard for floppies in the late 1980s. A 3.5 inch diskette measures about 3.5 inches wide by 3.75 inches deep by about .13 of an inch thick, and it is quite stiff. The term *floppy* comes from diskettes used on early personal computers; most were 5.25 inches wide and quite flexible. "High density" refers to the amount of information the diskette holds—about 1.44 megabytes of data.

Look carefully at a floppy diskette and you'll see an arrow indicating how you should insert the floppy into the disk drive. On the back of the diskette, you'll see a slide that you can move back and forth. Once you save a file to a diskette, you can lock the diskette so that you can no longer save information to that diskette or over write files. To lock the diskette, use your finger or a ball point pen to move the slide back and forth.

Floppy Drives The floppy disk drive is the device where you insert floppy diskettes for storing what you've written and where the computer "saves" or "writes" the information to a floppy diskette.

Freeware Freeware is copyrighted software that a programmer gives away or allows others to use at no cost. Some authors impose strict limitations on the use of the software they distribute as freeware, such as forbidding its resale and use in other programs.

Groupware Software that allows writers to work together. Lotus Notes is an example of groupware.

Hard Drive The hard drive, a permanent storage device, consists of electrical and mechanical components. A disk or series of small disks spin at thousands of revolutions per second and a magnetic head "writes" the information to the disk. Hard drive sizes are measured in megabytes and gigabytes. A byte represents one character or letter. A four gigabyte hard drive, for instance, can hold about four billion bytes or characters. Most computers will have one hard drive to hold the software and store files. As software programs and operating systems have grown more complex, they have grown larger, resulting in the need for larger hard drives. By the late 1990s, most new computers typically had hard drives of at least two gigabytes.

Internet The Internet is a network of computers connected to each other by telephone lines, high speed data lines, microwave relays, and satellite links. Sometimes referred to as the "network of networks," the Internet was developed by the United States government to support communication among computers in the event of a national catastrophe, such as a nuclear attack.

Keyboards The standard keyboard is often called the QWERTY keyboard, after the top left-hand row of letters. QWERTY keyboards have letter keys arranged like standard typewriters, additional function keys across the top and/or sides, and a numeric key pad on the right-hand side.

Machine Code Machine code is binary code (strings of ones and zeros) that tells the computer what to do in order to carry out your commands.

Macro A sequence of commands that are recorded and subsequently played back again. Writers can use macros to carry out tasks that have many steps, such as applying several types of formatting commands to a passage of text.

Modem A modem allows your computer to connect to the telephone system. Using modems, you can send faxes, make telephone calls, and connect to the Internet or commercial services, such as CompuServe or America Online. Internal modems are cards that slip into a slot on your computer's motherboard. External modems are housed in cases and connect to your computer via a connector on the back of the computer.

Modems come in different speeds, or baud rates; the higher the baud rate, the faster the modem can transfer information between computers. By the mid-1990s, manufacturers were selling computers with 28,800, 33,300, and 56,000 baud modems. A 56,000 baud modem transfers information about twice as fast as does a 28,800 baud modem—if your telephone system can handle the faster speed. Newer and still faster modems are being developed that can connect to cable television systems.

Monitors Computer monitors display information on a computer. Computer monitors may work similar to television sets—in other words, a phosphor-coated vacuum tube is "painted" by an electron gun. Or they may use light-emitting diodes (LEDs) such as those found on laptops and some newer computers. Color monitors are the norm, but you may on rare occasions encounter older monochrome monitors which usually display green, orange, or white letters, numbers, and symbols on a black background.

Most color monitors come in 12, 13, 15, 17 inch and even larger screen sizes. Monitors can display images at a range of resolutions. Resolution is measured in pixels, the unit of display. A typical resolution on both Apple Macintosh computers and IBM PC computers is 640 pixels wide by 480 pixels high. On PCs, this resolution is referred to as VGA resolution. Higher resolutions include 800 by 600 and 1,024 by 768. You can change a monitor's display resolution by using software provided by the maker of the video card or by setting options using the operating system.

Motherboard The motherboard consists of the plastic and fiber base with printed electrical circuits and slots for plugging in the CPU chip, random access memory, power supply, video graphics board, sound card, disk drives, and other components.

Mouse and Pointing Devices Pointing devices allow you to move the cursor around a screen without using the keyboard's arrow or other command keys. Pointing devices include the mouse, track ball, tablet, and touch pad. As you move a mouse or track ball, or as you move your finger across a touch pad, the cursor moves across the screen. If you hold the cursor in position and click on a button associated with the pointing device (e.g., a mouse button), you can invoke different functions for editing and writing, such as cutting, copying, pasting, and deleting text.

Network A system that allows computers to communicate with each other. Networks use wire, fiber optic cables, or radio signals to create connections among computers. Local Area Networks (LANS) usually connect computers in a small area such as a room or building. Wide Area Networks (WANS) usually connect several LANS in a larger area, such as a campus. The Internet is a network in the sense that it allows many computers to connect to each other. But it does not use a particular network operating system (NOS).

Network Cards Network cards allow your computer to connect to computer networks such as an Ethernet network at a school, government agency, or business. Network cards typically provide connections that are much faster than those provided by modems. Typically, however, you cannot access a telephone system using a network card.

Network Operating Systems (NOS) Network operating systems (NOS) provide software instructions that allow computers to communicate with a server or with other computers using the same operating system. To connect to a NOS, a computer needs a network card or modem and a connection to the system. Each computer's network card translates electronic signals and sends them over cabling that links the computers together. By using a networked computer, you can share files, use printers, send electronic mail, access other computers, the Internet, and the World Wide Web.

Typically, organizations that use network operating systems have a "systems administrator" who installs network cards and software and maintains the cabling and hardware that makes the network function properly. In most cases, you need not concern yourself with the complexities of the network.

Newsgroup An online discussion list that can be read with software called a Newsgroup Reader. A newsgroup consists of messages posted on a topic and organized by date or subject.

Operating Systems (OS) Operating systems enable you to manage the software on your computer and the documents you create. The operating system for IBM's first personal computers was DOS, disk operating system, written by Microsoft's Bill Gates, Paul Allen, and their fledgling staff. The first graphical operating system for personal computers was the Macintosh operating system. Later, Microsoft developed the Windows operating system which eventually was developed into Windows 3.x, Windows 95, Windows 98, and Windows NT.

Power Supply The power supply is an electronic part that divides and distributes power to the internal parts of the computer. Power supplies are often cooled by a fan. When you turn on many computers, you can hear the fan blowing. Look at the back of any computer and you will see the fan and exhaust vents, and you can feel the warm air being exhausted from inside the computer. Computer cases usually have holes along the front or sides that let cool air in.

Printers Printers allow you to create paper copies of your documents. The most common types of printers are laser printers, ink-jet printers, and dot matrix printers. Laser printers work much like copy machines. When you choose the PRINT command, your computer sends signals to the printer, and the printer deposits toner on paper to create the words and images. The printer then "sets" or "fixes" the toner with a heat process to ensure that the words and images won't rub off when you handle your printed document. Ink-jet printers spray the ink on the paper to create the words and images. Dot matrix printers use a ribbon that is impacted by small pins on the print head. Early printers used 9-pin heads, while later dot matrix printers used 28-pin heads. The larger number of pins produced a higher quality image.

Image quality is measured in dpi—dots per inch. The more dots per inch, the finer the level of detail the printer produces. Early laser printers printed at 300 dpi while newer models print at 600, 1,200, or higher dpi. In most cases the 300 to 600 dpi resolutions produce good quality images of your documents. Commercial printers use higher resolutions, typically 1,200 dpi or higher, for printing magazines, books, and other publications.

In many places, black and white printers have given way to color printing. Although standard black and white printers work fine for most writing, some organizations use color printers to add visual appeal to their documents.

Public Domain Software Software that has not been copyrighted; it is available to others without charge.

RAM Random access memory (RAM) is the temporary memory used by the computer. When you begin typing, words will appear on the screen. They're held in RAM until you save your writing (documents) to a floppy diskette, hard drive, network drive, or some sort of removable media. And save them you must. Should the electrical power fail and you haven't saved your words as a file, they're lost.

ROM Read only memory (ROM) holds the basic instructions that control your software and make your computer run.

Shareware Copyrighted software programs that you can use for a specified trial period. If you like the software, you pay the author a fee for your copy. You can often download shareware from the Internet.

Telnet Telnet is a software program that allows you to remotely control another computer on the Internet. To use Telnet, you need a Telnet program, access to the Internet from the computer you are using, and permission to use the remote computer. You'll also need to know the Internet name or address of the computer you want to access and the commands you will need to operate the remote computer's operating system and programs.

URL A uniform resource locator (URL) directs communications software to a particular file in a particular directory on a particular computer somewhere on the Internet.

Video Adapter Video adapters display text and images on a computer monitor. Some computers use display adapters that are built into the motherboard. Other computers use video display adapter cards that fit into a slot on the motherboard. The capabilities of your video display adapter determine, in part, the resolution at which you can view images on your computer monitor, the ability to view moving images, and the ability to view images in three dimensions.

Voice Recognition Software Software that enables you to talk to your computer. Voice recognition software allows you to dictate text into a word processing program and control the workings of your operating system and software programs. Although keyboards remain the primary way of entering information into computers, voice recognition software is rapidly improving and may replace keyboarding in the coming decades.

Word Processor Word processing software enables you to write quickly and efficiently on the computer without learning complex computer terminology and programming languages. As programmers developed word processing software, they added more and more functions to speed the writing process. You can typically purchase word processors as a stand-alone program or as part of an office suite, a selection of software packaged together usually including a word processing program, a spreadsheet program, a presentation program, and other programs. Although word processing programs have similar functions, the names of specific commands may vary. In most cases, if you learn to use one word processing program, you'll find it easy to learn another.

World Wide Web The World Wide Web, often referred to as simply the Web, is a subset of documents available on the Internet. Using the Web, you can easily view and move between documents, even when those documents are on computers literally thousands of miles apart. To visit a document, you need to know it's URL—uniform resource locator.

Web documents can contain formatted and unformatted text, graphics, sound, video, links to other documents, and even programs that allow you, for example, to calculate how much money you'll need to save to ensure a comfortable retirement. Web documents are viewed through browsers such as Netscape Navigator and Microsoft Internet Explorer.

appendix B

Word Processing Commands

BASIC EDITING COMMANDS

SELECT, MOVE, DELETE, CUT, COPY, and PASTE

Even the simplest word processing programs allow you to **SELECT, MOVE, DELETE, CUT, COPY,** and **PASTE** text. These commands are the heart of word processing.

SELECT: To select text means that you highlight text on your screen. Once selected, text can be deleted, cut, or copied. You can select text using the mouse (typically by clicking and/or dragging the mouse), the keyboard (typically by using the shift and arrow keys), or menu commands.

MOVE: To move text means to change the location of text by dragging it with your mouse from one position within a document to another. Most leading word processing programs support the **MOVE** command. To use it, click for a moment on the selected text until the cursor changes to indicate that the text can be moved and then drag it to its new position.

DELETE: To delete text means to remove it from the document without placing it into computer memory. Deleted text cannot be pasted into a document because it is not stored in computer memory. Delete text by hitting the **DELETE** or **BACKSPACE** key. In many word processing programs, you can also delete text by typing over it.

CUT: Cutting text involves selecting the text (using either a mouse or keyboard commands), removing it from the screen, and placing it into computer memory. The text, once cut, can be pasted into another part of the document or into a different document (see Panel 1.2). You can cut text by using the **EDIT** and then **CUT** commands (i.e., activate the **EDIT** menu using your mouse or the keyboard and select the **CUT** command; see below), by using a keyboard command (such as Ctrl-x or Command-x), or by clicking with your mouse on a **CUT** icon in a button bar.

COPY: Copying text involves selecting the text and placing a duplicate of the text into computer memory. Once copied, text can be pasted into another part of the document or into a different document. You can use menu commands, keyboard commands, and button bar icons to copy text.

PASTE: Pasting text involves placing text that is held in computer memory into a document. Text held in computer memory can be pasted repeatedly into a document or documents. You can use menu commands, keyboard commands, and button bar icons to paste text.

UNDO, REDO, and REPEAT

Writers can change their minds. When you are working with a word processing program, the **UNDO** and **REDO** commands allow you to reverse a previous decision. If you delete some text, then decide you shouldn't have, use the **UNDO** command to restore the text. Depending on your word processing software, you can either **UNDO** a single action or a list of actions. The latter option is called **MULTIPLE UNDO** or **UNLIMITED UNDO.** In some word processing programs, you can **UNDO** everything you've done to a file since you opened it.

The **REDO** command allows you to reverse an **UNDO** command. If you **UNDO** something, but then decide that you should have left it as it was, use the **REDO** command. Over time, you'll find the **UNDO** and **REDO** commands highly useful.

The **REPEAT** command allows you to do the same action over and over again. You'll find the **REPEAT** command useful, for instance, if you want to apply the same formatting commands to several paragraphs in a document. Rather than using several mouse clicks, keyboard commands, or menu selections to format each paragraph, use the **REPEAT** command to simplify the process.

Like **CUT, COPY,** and **PASTE,** you can invoke the **UNDO, REDO,** and **REPEAT** commands by using menus, keyboard commands, and button bars.

KEYSTROKE COMMANDS

Many Windows-based software programs use similar keystroke commands to carry out the same functions. Table B.1 provides a list of these commands.

TABLE B.1 Editing keystrokes common to word processing programs

Task	Keystrokes/Keys PC	Macintosh
Cut or delete highlighted text	Ctrl + x	⌘ + x
Copy highlighted text	Ctrl + c	⌘ + c
Paste highlighted text	Ctrl + v	⌘ + v
Clear	Delete	Del
Insert copy	Insert	Ins
Move up a screen	Page Up	Page Up
Move down a screen	Page Down	Page Down

FILE FUNCTIONS ON WINDOWS WORD PROCESSORS

Look carefully at the menus on each word processing program. Compare the File menus below of Microsoft Word and Corel WordPerfect, and you'll find each offers similar functions. Figure B.1 shows the FILE commands from the Corel WordPerfect 8 file menu, while Figure B.2 shows the FILE commands from the Microsoft Word 97 menu. Once you've opened Microsoft Word and Corel WordPerfect on a Macintosh, the word processors, for the most part, use similar menu commands.

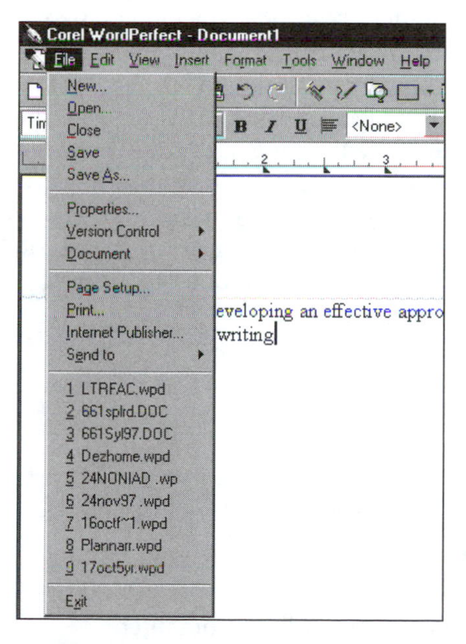

FIGURE B.1 Corel WordPerfect 8 file menu

FIGURE B.2 Microsoft Word 97 file menu

appendix C

Writing with Older Computers

If you're short on cash and want a computer for basic word processing, you can purchase an older computer for $50 to $500, and the sellers will often include the software, manuals, printer, and modem. In some cases, people who have older computers are willing to give them away for free. Try to find a complete system so that you do not need to install a printer and printer driver (software that tells the word processing program what software is available). Many older computers are based on the DOS, Windows 3.1x, Apple II, or older Macintosh operating systems.

Although older computers will work well for writing, keep in mind the following limitations: (1) they'll be slower than current computers, (2) you may find the operating systems and software more difficult to learn than Windows and Macintosh systems, (3) you may have trouble transferring files from an older computer to newer machines, and (4) the word processing software will have fewer functions than the most recent word processing programs.

If you're considering buying an older machine, find a relative, friend, or someone familiar with older machines who can check it out. To inspect a machine that you are thinking about purchasing, be sure the

- Computer and monitor turns on
- Disk drives work
- Computer can print to a printer
- Computer can save to a diskette
- Computer hard disk has sufficient space to save files
- Word processing software opens and functions
- Manuals come with the computer
- Keyboard templates are available
- Original software diskettes or CD-ROMs come with the computer
- Computer and monitor are clean

As you assess the computer, listen for any unusual noises or sounds coming from the fans, disk drives, or CD-ROM drive, if it is so equipped. Smell the machine. Avoid any computer, or any electrical equipment for that matter, that smells "hot" or unusual.

If you're concerned about compatibility issues, such as transferring files from one computer to another, keep in mind that it is sometimes difficult to transfer files from an older computer to a newer computer.

Because fewer and fewer computers are equipped with the older 5.25 inch floppy diskettes, look for used computers that have 3.5 inch high density floppy diskettes. Make sure that it's a *high* density diskette—some older machines have 3.5 inch *double* density floppy disk drives. Although transferring files between high and double density 3.5 inch floppy diskettes is possible, it may not work in some cases.

If you'll be writing on the older computer, saving files to a diskette, and opening the files on a computer in a laboratory or office, make sure the older computer is compatible with the laboratory or office machines. To begin, create a file on the lab or office computer, use **SAVE AS** and save the file as in ASCII or text format to the floppy drive. Take that diskette along with you when you inspect the new computer.

Check to see if the used computer can read the diskette from its operating system and save to the floppy drive. Open the word processing program on the used computer and try to open the ASCII or text file. In most cases, it should open. If it doesn't, you may not be able to transfer files back and forth between the used computer you're considering and the office or laboratory computer.

While you're trying the computer, write a short passage and save it to the computer's hard drive. Close the file, open it again, and try to print it. If you or your friend have any problems, consider whether or not you'll want to work with the computer to solve the problems.

Many earlier versions of word processing programs came with a keyboard template that displayed the key strokes used to carry out commands. Although learning the key stroke combinations for these commands requires some time, you seldom needs more than a couple of dozen commands for most writing tasks like opening, saving, printing, editing, spell checking, formatting, and printing.

Try to find a computer–word processor–printer system that is completely installed. For instance, it can be difficult to locate a printer driver (software) for an older computer and printer. Obtaining the printer driver may be time-consuming or costly. In some cases, you may not be able to find a printer driver for the word processing software that works with the respective printer, word processing program, and computer. Second, installing different printers on older machines and adjusting the printing functions can be time-consuming and difficult. Even people knowledgeable about printers may need a couple of hours to set up the system.

Make sure that ribbons, cartridges, or other printer supplies are available for the system you'll be buying. In most cases, computer supply catalogs list a wide range of ribbons, cartridges, and refills for printers.

If you're purchasing an older computer that runs a DOS operating system, a few basic commands should be helpful (Table C.1). The commands are printed in Courier. Note too that the commands are followed by pressing the ENTER key. While Table C.1 provides a short list of commands, check the DOS manual that accompanies the computer for a detailed list of DOS commands. In some cases, the DOS manuals were installed on computers with hard drives, so check there for the information as well.

If you can't locate a manual, check your local library for books on the basics of using DOS operating systems and earlier computers. One especially useful book for learning the DOS commands is Van Wolverton's *Running MS DOS*, published by Microsoft Press with several editions in the 1980s.

TABLE C.1 Selected DOS commands for computers with DOS operating systems

Tasks	Command	Typing Commands	Notes
Copying one file at a time	Copy	Copy A:Filename B:	A: B: and C: represent the disk drives
Copying all files from a subdirectory	Copy *.* A:	Copy *.* A: Press Enter Key	Command copies all files from the current subdirectory to the floppy diskette.
Making directories and subdirectories	Mkdir	Mkdir Old Press Enter Key	Old represents the name of the directory
Changing Drives	CD...	CD... Press Enter Key	Type CD, three spaced periods, and then press the ENTER key.
Listing files in a directory	Dir	Dir Press Enter Key	Type Dir and then press the ENTER key.
Removing or erasing files	Erase	Erase <filename> Press Enter Key	Type Erase, the name of the file, and then press the ENTER key.
To learn how much space is left on a disk	Dir	Dir Press Enter Key	The computer examines the directory, subdirectory, or floppy diskette and reports the number of files and the available space on the disk drive or floppy diskette.

appendix D

Reference Works and Style Guides

STYLE MANUALS

Achtert, W. S., & Gibaldi, J. (1985). *The MLA style manual.* New York: Modern Language Association.

Achtert, W. S., & Gibaldi, J. (1988). *The MLA handbook for writers of research papers* (3rd ed.). New York: Modern Language Association.

American Psychological Association. (1994). *Publication manual of the American Psychological Association* (4th. ed.). Washington, DC: Author.

Associated Press Staff. (1987). *The Associated Press stylebook and libel manual: The journalist's bible.* Reading, MA: Addison-Wesley.

CBE Style Manual Committee. (1995). *CBE style manual: A guide for authors, editors, and publishers in the biological sciences.* (6th ed.). Bethesda, MD: Council of Biology Editors.

Chicago Editorial Staff. (1993). *The Chicago manual of style.* (14th ed.). Chicago: University of Chicago Press.

Dodd, J. S. (1997). *The ACS style guide: A manual for authors and editors.* Washington, DC: American Chemical Society.

Hurth, E. J. (1987). *Medical style and format: An international manual for authors, editors and publishers.* Philadelphia, PA: Institute for Scientific Information.

Iverson, C. (1997). *American medical association manual of style: A guide for authors & editors.* Baltimore: Williams & Wilkins.

Jordon, L. (1982). *The New York Times manual of style and usage.* New York: Random House.

Lane, M. K., Lindenfelser, L. F., & Bingham, P. G. (1993). *Style manual for political science.* Washington, DC: American Political Science Association.

Skillin, M. E., Gay, R. M., & others. (1974). *Words into type.* Upper Saddle River, NJ: Prentice Hall.
U.S. Government Printing Office style manual. (1984). Washington, DC: US Government Printing Office.
Webster's standard American style manual. (1985). Springfield, MA: Merriam-Webster.

GRAMMAR AND USAGE GUIDES

Bernstein, T. M. (1978). *The careful writer.* New York: Simon & Schuster.
Carter, B., & Skates, C. (1996). *The Rinehart handbook for writers* (4th ed.). Fort Worth, TX: Harcourt Brace College.
Crews, F. (1992). *The Random House handbook* (4th ed.). New York: McGraw.
Fulwiler, T., Hayakawa, A. R., & Kupper, C. (1995). *The college writer's reference.* Upper Saddle River, NJ: Prentice Hall.
Hacker, D. (1996). *Rules for writers* (3rd ed.). Boston: Bedford Books.
Hacker, D. (1996). *Pocket style manual.* Boston: Bedford Books.
Hodges, J. C., Horner, W. B., Webb, S. S., & Miller, R. K. (1997). *Harbrace college handbook,* (13th ed.). Fort Worth: Harcourt Brace Jovanovich.
Kessler, K., & McDonald, D. (1996). *When words collide* (4th ed.). Belmont, CA: Wadsworth.
Kriszner, L. G., & Mandell, S. R. (1998). *The brief Holt handbook* (2nd ed.). Fort Worth: Harcourt Brace College.
Plotnik, A. (1982). *The elements of editing.* New York: Macmillan.
Troyka, L. Q. (1996). *Simon & Schuster handbook for writers.* Upper Saddle River, NJ: Prentice Hall.

DICTIONARIES

The American heritage dictionary of the English language (3rd ed.). (1994). Boston: Houghton Mifflin.
The American heritage dictionary: College edition (3rd ed.). (1994). Boston: Houghton Mifflin.
The American heritage Stedman's medical dictionary. (1995). Boston: Houghton Mifflin.
Berstein, T. M. (1988). *Berstein's reverse dictionary.* New York: Random House.
Dictionary of computer words. (1995). Boston: Houghton Mifflin.
Elsevier's dictionary of geosciences. (1991). New York: Elsevier.
Elsevier's dictionary of computer science and mathematics. (1995). New York: Elsevier.
Karush, W. (1989). *Webster's world dictionary of mathematics.* Upper Saddle River, NJ: Prentice Hall.

McGraw-Hill dictionary of earth sciences. (1996). New York: McGraw-Hill.
McGraw-Hill dictionary of scientific and technical terms. (5th. ed.). (1995). New York: McGraw-Hill.
Merriam Webster's collegiate dictionary (10th ed. Rev.). (1996). Springfield, MA: Merriam-Webster.
Phaffenberger, B. (1997). *Dictionary of computer terms.* New York: Macmillan.
The concise Oxford dictionary of earth sciences. (1990). New York: Oxford University Press.
The Random House dictionary of the English language. (1995). New York: Random House.
Webster's new world college dictionary. (1989). New York: Macmillan.
Webster's new world dictionary and thesaurus. (1996). New York: Macmillan.
Webster's new world dictionary of media and communications. (1995). New York: Macmillan.
Webster's unabridged dictionary. (1997). New York: Random House.

BOOKS ON WRITING AND INFORMATION GATHERING

Anderson, D., Benjamin, B., & Paredes-Holt, B. (1998). *Connections: A guide to online-writing.* Boston: Allyn and Bacon.
Baker, S. (1998). *The practical stylist.* (8th ed.). New York: Longman.
Beer, D., & McMurrey, D. (1997). *A guide to writing as an engineer.* New York: John Wiley.
Brusaw, C. T., Alred, G. J., & Oliu, W. E. (1997). *Handbook of technical writing* (5th ed.). New York: St. Martin's Press.
Brusaw, C. T., Alred, G. J. & Oliu, W. E. (1997). *The business writer's handbook.* New York: St. Martin's Press.
DeVries, M. A. (1992). *Prentice Hall manual: A complete guide with model formats for every business writing.* Upper Saddle River, NJ: Prentice Hall.
Kessler, L., & McDonald, D. (1996). *When words collide* (4th ed.). Belmont, CA: Wadsworth.
Michaelson, H. (1996). *How to write and publish engineering papers and reports* (3rd. ed.). Phoenix, AZ: Oryx.
River, W. L., & Rodriguez, A. W. (1995). *A journalist's guide to grammar and style.* Boston: Allyn and Bacon.
Rodrigues, D. (1997). *The research paper and the World Wide Web.* Upper Saddle River, NJ: Prentice Hall.
Vitanza, V. J. (1998). *Writing for the World Wide Web.* Boston: Allyn and Bacon.
Zimmerman, D. E., & Muraski, M. (1995). *Elements of information gathering.* Phoenix, AZ: Oryx.
Zimmerman, D. E., & Rodrigues, D. (1992). *Research and writing in the disciplines.* Fort Worth, TX: Harcourt Brace Jovanovich College Publishers.
Zinnser, W. (1996). *On writing well* (5th ed.). New York: Harper and Row.

REFERENCES

Anderson, B., & Anderson, J. (1997). *Computer & desk stretches.* Palmer Lake, CO: Stretching. P.O. Box 767, Palmer Lake, CO 80133.

Cassingham, R. C. (1986). *The Dvorak keyboard.* Boulder, CO: Freelance Communications. P.O. Box 17326, Boulder, CO 80308.

Covington, M. A. (1992). *Dictionary of computer terms.* Hauppauge, NY: Barron's.

Dvorak, A. (1936). *Typewriting behavior.* New York: American Book Company. (Available from Freelance Communications, P.O. Box 17326, Boulder, CO 80308-0326).

Felkner, D. B., Pickering, F., Charrow, V. R., Holland, V. M., & Redish, J. C. (1982). *Guidelines for document designers.* Washington, DC: American Institutes for Research.

Freedman, A. (1995). *The computer glossary: The complete illustrated dictionary.* New York: Amacom.

Furger, R. (1993). Weird keyboards; Rx for hands? *PC World, 11*(5), 118–124.

Hartley, J. (1985). *Designing instructional text.* New York: Kogan Page, Nichols.

Hill, M., & Cochran, W. (1977). *Into print.* Los Altos, CA: William Kaufman.

Leavitt, S. B. (1995). *Vision comfort at VDTs.* Irwindale, CA: MicroCentre™. Available on request. Telephone 1-800-966-5511.

Linden, P. (1995). *Compute in comfort.* Upper Saddle River, NJ: Prentice Hall.

Maran, R. (1996). *Computers simplified* (3rd ed.). Foster City, CA: IDG Books.

Microsoft. (1994). *Microsoft natural keyboard.* Redmond, WA: Microsoft.

Robinson, P. (1994). William F. Buckley Jr. *Forbes ASAP,* October 10, 1994, 60–69.

Schriver, K. (1995). *Dynamics in document design.* New York: John Wiley.

Sellers, D. (1994). *Zap! How your computer can hurt you—and what you can do about it.* Berkeley, CA: Peachpit.

Sellers, D. (1995). *25 steps to safe computing.* Berkeley, CA: Peachpit.

Tinker, M. A. (1963). *Legibility of print.* Ames, IA: Iowa State University Press.

Index

Credits:

Screen image on page 113, reproduced with the permission of Digital Equipment Corporation. AltaVista, the AltaVista logo, and the Digital logo are trademarks of Digital Equipment Corporation.

Screen images on pages 35, 86, 90, 150, 173, System Software 8.1 Copyright © Apple Computer, Inc. Used with permission. Appler and the Apple logo are registered trademarks of Apple® Computer, Inc. All rights reserved.

Screen images on pages 96, 98, reprinted by permission from CARL/UNCOVER.

Screen images on pages 6, 11, 12, 33, 37, 38, 48, 57, 75, 148, 155, 158, 247, reprinted by permission from Corel Corporation.

Screen image on page 117, Electric Library is a registered trademark of Infonautics, Inc. or its subsidiaries. The Electric Library service is Copyright © 1998 Infonautics Corporation. All rights reserved.

Screen image on page 114, Excite, Excite Search, and the Excite Logo are trademarks of Excite, Inc. and may be registered in various jurisdictions. Excite screen display Copyright © 1995–1997 Excite, Inc.

Screen images of FirstSearch Catalog on pages 111, 112, reprinted with permission of the OCLC Online Computer Library Center, Inc. FirstSearch is a trademark of OCLC Online Computer Library Center, Inc.

Screen image on page 100, Lamar screen reprinted by permission of ACNS, Colorado State University.

Screen image on page 118, Copyright © 1997 Lycos, Inc. Lycos® is a registered trademark of Carnegie Mellon University.

Screen images on pages 7, 8, 9, 12, 21, 22, 23, 32, 47, 51, 54, 76, 81, 89, 95, 98, 102, 104, 132, 138, 146, 147, 149, 152, 156, 160, 161, 164, 174, 175, 176, 177, 178, 179, 180, 181, 182, 186, 225, 226, 247, reprinted by permission from Microsoft Corporation.

Screen images by Netscape. Netscape Communications Corporation has not authorized, sponsored, or endorsed, or approved this publication and is not responsible for its content. Netscape and the Netscape Communications Corporate Logos, are trademarks and trade names of Netscape Communications Corporation. All other product names and/or logos are trademarks of their respective owners.

Screen images of SAGE on page 95, reprinted by permission of Colorado State University Libraries. Colorado State University Libraries uses Innovative Interfaces, Inc. as a platform for its online catalog.

Photographs on pages 208, 209, 214, 221, courtesy of Thomas M. Schaefges/Omegatype Typography, Inc.

Screen images of gopher sites on pages 102, 103, reprinted by permission of the University of Minnesota Office of Information Technology.